MATHEMATICAL PROBLEM SOLVING

Yearbook 2009

Association of Mathematics Educators

MATHEMATICAL PROBLEM SOLVING

Yearbook 2009

Association of Mathematics Educators

Editors

Berinderjeet Kaur • Yeap Ban Har • Manu Kapur

National Institute of Education, Singapore

AME
ASSOCIATION OF MATHEMATICS EDUCATORS

World Scientific

NEW JERSEY • LONDON • SINGAPORE • BEIJING • SHANGHAI • HONG KONG • TAIPEI • CHENNAI

Published by

World Scientific Publishing Co. Pte. Ltd.
5 Toh Tuck Link, Singapore 596224
USA office: 27 Warren Street, Suite 401-402, Hackensack, NJ 07601
UK office: 57 Shelton Street, Covent Garden, London WC2H 9HE

British Library Cataloguing-in-Publication Data
A catalogue record for this book is available from the British Library.

Cover photo from Princess Elizabeth Primary School, Singapore (2008).

MATHEMATICAL PROBLEM SOLVING
Yearbook 2009, Association of Mathematics Educators

ISBN-13 978-981-4277-20-4
ISBN-10 981-4277-20-7
ISBN-13 978-981-4277-21-1 (pbk)
ISBN-10 981-4277-21-5 (pbk)

Contents

Part I

Introduction

Chapter 1

Mathematical Problem Solving in Singapore Schools

Berinderjeet KAUR YEAP Ban Har

This opening chapter provides a view of the development of mathematical problem solving in Singapore schools. From a research and curriculum development perspective, this chapter shows how research and development elsewhere had impacted upon the emergence and subsequent development of mathematical problem solving in Singapore schools. From a pedagogical perspective, the chapter shows the range of problem-solving processes students engage in, the variety of pedagogy options available to teachers and the array of tasks that can bring the processes and pedagogy together. From an assessment perspective, the chapter suggests how tasks used in national examinations have a direct influence on the implementation of a problem-solving curriculum. From an economic perspective, this chapter argues that an effective implementation of a problem-solving curriculum equips students with the necessary competencies for a knowledge-based economy.

1 Introduction

In 1992 mathematical problem solving was made the primary goal of the school mathematics curriculum in Singapore. Since then, though the curriculum has been revised twice, in 2001 and 2007, mathematical problem solving has remained its primary goal. Figure 1 shows the mathematics curriculum framework for Singapore schools (Ministry of Education, 2006a, 2006b). The emphasis on mathematical problem solving was influenced by recommendations in documents such as An Agenda for Action (National Council of Teachers of Mathematics, 1980) and the Cockcroft Report (Cockcroft, 1982) from the United States and the United Kingdom respectively. Today, it is rare to find a mathematics curriculum that does not place emphasis on mathematical problem solving.

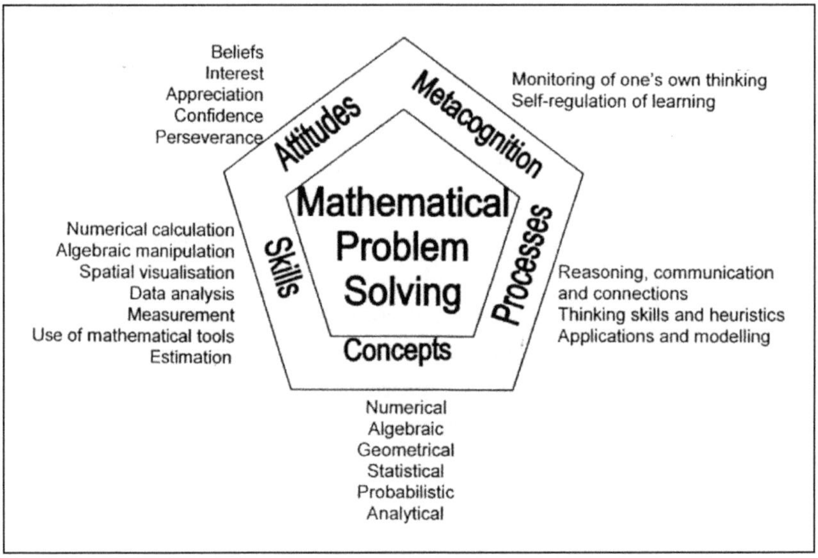

Figure 1. Framework of the Singapore school mathematics curriculum

The seminal doctoral work of Kilpatrick (1967) involving the analysis of solutions of word problems in mathematics at Stanford University and subsequent work by himself and other researchers have established mathematical problem solving as a research field. In particular, Kilpatrick's (1978) classic paper, Variables and Methodologies in Research on Problem Solving, outlined key research variables in the field. Since then, mathematical problem solving as a research field has grown and matured to some extent (Lester, 1994; Lesh & Zawojewski, 2007). This has certainly been the case in Singapore

(Foong, 2009). In a state-of-the art review in the early 1990s, Chong, Khoo, Foong, Kaur and Lim-Teo (1991) found that research in mathematics education in Singapore, in general, and problem solving, in particular, to be in its state of infancy. Since then, significant work had been done. Early studies in mathematical problem solving on students (Kaur, 1995) and teachers (Foong, 1990) have stimulated further research into the domain. Kaur (1995) investigated the strategies used by middle school students in solving non-routine problems and clarified the relationship between students' ability to perform particular mathematical procedures and their ability to solve problems. Foong (1990) investigated the problem-solving processes used by pre-service teachers in solving non-routine problems. A recent review of research, by Foong (2009), on mathematical problem solving in Singapore has indicated that our knowledge on problem-solving approaches and tasks used in the classroom, teachers' beliefs and practices, and students' problem-solving behaviours have grown. It is important that such rich research findings find their way into the classrooms. This book showcases several research findings and theories translated into classroom practice.

2 Mathematical Problem Solving

Mathematical problem solving occurs when a task provides some blockage (Kroll & Miller, 1993). Lester (1983) describes a mathematical problem as a task that a person or a group of persons want or need to find a solution for and for which they do not have a readily accessible procedure that guarantees or completely determines the solution.

How does the mathematics textbooks used in Singapore encourage problem solving? Ng (2002) found that the majority of the problems in the primary textbooks were word problems that are closed and routine. Open-ended problems were not common. Fan and Zhu (2000) found that while the lower secondary textbooks provided students with a strong foundation in problem solving, more open-ended problems as well as authentic real-life problems could be included. It is, thus, timely that several chapters in this book attempts to broaden the conception of what it means to engage in mathematical problem solving. The chapter by **Yeo Kai Kow** describes the importance of open-ended problems in lower secondary levels. The chapter by **Yeo Boon Wooi** and **Yeap Ban Har** clarifies the relationship between mathematical problem solving and mathematical investigation. The chapter by **Ang Keng Cheng** helps readers understand the role of mathematical modeling in real-world mathematical problem solving. **Yeap Ban Har** described the processes in mathematical problem posing to show its relationship to mathematical problem solving.

3 Pedagogy and Practice in Mathematical Problem Solving

Textbook analysis studies and classroom studies have shown that the vast majority of textbook tasks are well-structured tasks (Ng, 2002; Fan & Zhu, 2000) and classroom instruction is mostly teacher-led (Ho, 2007). Foong (2002) has found that teachers in Singapore tend to adopt the teaching for problem solving approach where the emphasis is learning mathematics content for the purpose of applying them to a wide range of situations. Ho's (2007) case studies of four primary-level teachers confirmed, and provided more information for, this finding. With the call for a wider repertoire of teaching methods, in general, and of problem-solving instruction, in particular, it is necessary for teachers to explore alternative pedagogies for mathematical problem-solving instruction.

In the chapter by **Manu Kapur**, it is interesting to note that the use of ill-structured problems as well as students experiencing productive failure resulted in students performing significantly better in problem-solving tasks. The chapter by **Lillie Albert**, **Christopher Bowen** and **Jessica Tansey** describes note taking as a pedagogical tool to develop mathematical problem solving. The chapter by **Yoshinori Shimizu** provides an insider's perspective to the findings from an international study about the way mathematics lessons are conducted in typical Japanese classrooms and describes a typical mathematics lesson in Japan that is best described as structured problem solving. In the chapter by

Yeap Ban Har, how mathematical problem posing was used in several primary-level classes in Singapore is described.

With advances in information and communication technology, it is not possible to avoid the impact of technology on mathematical problem solving. Chua (2001) described the processes of social construction of mathematical ideas as students solved problems in pairs in a computer-mediated environment. In this book, the chapter by **Sarah Davis** shows the immense potential of a technology-supported classroom pedagogy that requires students to work together. The chapter by **Ang Keng Cheng** also emphasizes the central role of technology in mathematical modeling processes.

These chapters show how teachers in Singapore and elsewhere used pedagogy that departs from typical well-structured tasks and teacher-led classroom instruction. Such pedagogical practices provide readers with a repertoire of instructional models to teach mathematical problem solving in their own classrooms. The chapter by **Peter Sullivan**, **Judith Mousley** and **Robyn Jorgensen** provides research-based teacher actions that can facilitate mathematical problem solving.

4 Mathematical Problem-Solving Tasks

The Singapore mathematics curriculum defines problems to include a wide range of situations, including non-routine, open-ended and real-world problems (Ministry of Education, 2006a, 2006b). Figures 2, 3 and 4, show problems that students had to solve in the national examinations of recent years. The problem in Figure 2 was from the sixth grade national examination (Primary School Leaving Examination). The problem in Figure 3 was from the tenth grade national examination (General Certificate of Education Ordinary Level Examination). The problem in Figure 4 was from the twelfth grade national examination (General Certificate of Education Advanced Level Examination). Each of the problems was novel in that it was the only time a task of that type was posed in the respective examinations.

Table 1

1	2	3	4	5	6	7	8
9	10	11	12	13	14	15	16
17	18	19	20	21	22	23	24
25	26	27	28	29	30	31	32
33	34	35	36	37	38	39	40
41	42	43	44	45	46	47	48
49	50	51	52	53	54	55	56

Table 1 consists of numbers from 1 to 56. Kay and Lin are given a plastic frame that covers exactly 9 squares of Table 1 with the centre square darkened.

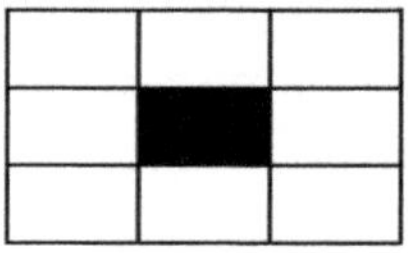

(a) Kay puts the frame on 9 squares as shown in the figure below.

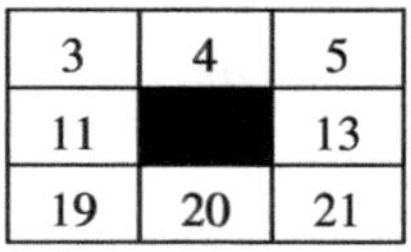

What is the average of the 8 numbers that can be seen in the frame?

(b) Lin puts the frame on some other 9 squares.
The sum of the 8 numbers that can be seen in the frame is 272.
What is the largest number that can be seen in the frame.

Figure 2. A problem from the grade six national examination (Singapore Examination and Assessment Board, 2009)

A fly, F, starts at a point with position vector (**i** + 12**j**) cm and crawls across the surface with a velocity of (3**i** + 2**j**) cm s^{-1}. At the instant the fly starts crawling, a spider, S, at the point with position vector (85**i** + 5**j**) cm, sets off across the surface with a velocity of (-5**i** + k**j**) cm s^{-1}, where k is a constant. Given that the spider catches the fly, calculate the value of k.

Figure 3. A problem from the grade 10 national examination (Ministry of Education, 2007)

Four friends buy three different kinds of fruit in the market. When they get home they cannot remember the individual prices per kilogram, but three of them can remember the total amount that they each paid. The weights of fruit and the total amounts paid are shown in the following table.

	Suresh	Fandi	Cindy	Lee Lian
Pineapple (kg)	1.15	1.20	2.15	1.30
Mangoes (kg)	0.60	0.45	0.90	0.25
Lychees (kg)	0.55	0.30	0.65	0.50
Total amount paid in $	8.28	6.84	13.05	

Assuming that, for each variety of fruit, the price per kilogram paid by each of the three friends is the same, calculate the total amount that Lee Lian paid.

Figure 4. A problem from the grade 12 national examination (Singapore Examination and Assessment Board, 2008)

Given that test items in Singapore's national examinations comprises of some problems, it is a challenge for teachers to generate such novel tasks for their students to attempt during instruction. The chapter by **Dindyal Jaguthsing** describes problems for secondary level-students and the processes students engage in when attempting them. The chapter by **Toh Tin Lam** shows tasks that have the ability to spark the curiosity in students. **Yeo Kai Kow** presents open-ended tasks that require students to delve into their conceptual understanding. **Catherine Vistro-Yu** shows how a familiar task can be systematically transform to generate a set of related tasks, some of which are novel. This technique is useful to Singapore teachers who often need to design worksheets comprising of a set of problems for students to consolidate their mathematical problem-solving ability. In the chapter by **Yoshinori Shimizu** readers are able to see how good lessons can be constructed around carefully-selected problems. The use of a set of related problems as well as centering lessons around good problems give students opportunities to have prolonged and deep engagement with the tasks.

5 Mathematical Problem Solving and the Education System in Singapore

The vision of the Ministry of Education in Singapore is Moulding the Future of the Nation i.e. education is perceived as critical to the survival of the country. Mathematics and other school subjects are platforms for students to develop a set of competencies that hold them in good stead to function well in the type of economy that Singapore engages in. It is no wonder that the Ministry of Education has over the years introduced a slew of initiatives, two of which are Thinking School, Learning Nation (TSLN) and Teach Less, Learn More (TLLM). TSLN aims to develop good thinking through school subjects. TLLM encourages teachers to reduce the content taught via direct teaching but instead engage students in meaningful activities so that they use knowledge to solve problems and whilst solving problems extend their knowledge through inquiry. Thus, a shift in the emphasis of mathematics teaching and learning from acquisition of skills to "development and improvement of a person's intellectual competence" (p.5, Ministry of Education, 2006a), makes it necessary for mathematics education to make mathematical problem solving and its instruction its focus. It is the aim of this book to provide readers with a range of ideas on how this can happen in the mathematics classroom.

6 Concluding Remarks

It has been 17 years since mathematical problem solving was introduced as the primary aim of learning mathematics in Singapore schools. While many teachers are now familiar with the notion of mathematical problem solving as well as various problem-solving heuristics used during problem solving, the challenge of balancing between developing fluent basic skills and problem-solving ability remains. Some teachers may perceive these as mutually exclusive. There are several chapters in this book that provide the alternate perspectives that acquisition of basics is not mutually exclusive with the development of mathematical problem-solving ability. Given that teachers are already familiar with the notion of mathematical problem solving, it is timely to step back and examine what it means to learn mathematics, and in the process, derive implications for mathematics education research and practice as well as some of the critical issues that the AME yearbooks could focus on in the coming years. Chapter 14 by **Manu Kapur** aims to do precisely this. By drawing on the folk categories of "learning about" a discipline and "learning to be" a member of the discipline (Thomas & Brown, 2007), Kapur proposes a move beyond the *pedagogy* of mathematics to include the *epistemology* of mathematics. To this end, he puts forth three essential research thrusts: a) understanding children's inventive and constructive resources, b) designing formal and informal learning environments to build upon these resources, and c) developing teacher capacity to drive and support such change.

Several chapters in this book arose out of the keynote lectures and workshops conducted during the annual Mathematics Teachers Conference of 2008 which was jointly organized by the Association of Mathematics Educators in Singapore and the Mathematics and Mathematics Academic Group at the National Institute of Education in Singapore. The annual conference is very well attended by mathematics teachers in Singapore with an increasing number of foreign teachers joining the event each year. The yearbook, of which this is the first in the series, provides multiple perspectives to a selected aspect of mathematics education – mathematical problem solving. Such a treatment of mathematical problem solving is done with a purpose of bring mathematical problem-solving instruction to the next level.

References

Chong, T. H., Khoo, P. S., Foong, P. Y., Kaur, B., & Lim-Teo, S. K. (1991). *A state-of-the-art review of mathematics education in Singapore.* Singapore: Institute of Education.

Chua, G. K. (2001). *A qualitative case study on the social construction of ideas in mathematical problem solving*. Unpublished dissertation, Nanyang Technological University, Singapore.

Cockcroft, W. H. (1982). *Mathematics counts: Report of the committee of inquiry into the teaching of mathematics in primary and secondary schools in England and Wales.* London: HMSO.

Fan, L. H. & Zhu, Y. (2007). Problem solving in Singapore secondary mathematics textbooks. *The Mathematics Educator, 5*(1/2), 117-141.

Ho, K. F. (2007). *Enactment of Singapore's mathematical problem-solving curriculum in Primary 5 classrooms: Case studies of four teachers' practices.* Unpublished doctoral dissertation, Nanyang Technological University, Singapore.

Foong, P. Y. (1990). *A metacognitive heuristic approach to mathematical problem solving.* Unpublished doctoral dissertation, Monash University, Australia.

Foong, P. Y. (2002). Roles of problems to enhance pedagogical practices in the Singapore classrooms. *The Mathematics Educator, 6*(2), 15-31.

Foong, P. Y. (2009). Review of research on mathematical problem solving in Singapore. In K. Y. Wong, P. Y. Lee, B. Kaur, P. Y. Foong & S. F. Ng (Eds), *Mathematics education: The Singapore journey* (pp. 263-300). Singapore: World Scientific.

Kaur, B. (1995). *An investigation of children's knowledge and strategies in mathematical problem solving.* Unpublished doctoral dissertation, Monash University, Australia.

Kilpatrick, J. (1967). Problem solving in mathematics. *Review of Educational Research, 39*, 523-534.

Kilpatrick, J. (1978). Variables and methodologies in research on problem solving. In L. L. Hatfield & D. A. Bradfard (Eds.), *Mathematical problem solving: Papers from a research workshop* (pp. 7-20). Columbus, OH: ERIC/SMEAC.

Kroll, D. L. & Miller, T. (1993). Insights from research on mathematical problem solving in the middle grades. In D. T. Owens (Ed.), *Research ideas for the classroom: Middle grades mathematics* (pp. 58-77). New York: Macmillan Publishing Company.

Lesh, R. & Zawojewski, J. (2007). Problem solving and modeling. In F. K. Lester (Ed.), *Second handbook on research on mathematics teaching and learning* (pp. 763-804). Charlotte, NC: Information Age Publishing and National Council of Teachers of Mathematics.

Lester, F. K. (1983). Trends and issues in mathematical problem-solving research. In R. Lesh & M. Landau (Eds.), *Acquisition of mathematical concepts and processes* (pp. 229-261). Orlando, FL: Academic Press.

Lester, F. K. (1994). Musings about mathematical problem-solving research: 1970-1994. *Journal for Research in Mathematics Education, 25*(6), 660-675.

Ministry of Education. (2006a). *Mathematics syllabus: Primary.* Singapore: Curriculum Planning and Development Division.

Ministry of Education. (2006b). *Mathematics syllabus: Secondary.* Singapore: Curriculum Planning and Development Division.

Ministry of Education. (2007). *Past Year Examination Questions 1996-2006: Additional Mathematics.* Singapore: Dyna Publishers.

National Council of Teachers of Mathematics (1980). *An agenda for action.* Reston, VA: Author.

Ng, L. E. (2002). *Representation of problem solving in Singaporean primary mathematics textbooks with respect to types, Polya's model and heuristics.* Unpublished MEd dissertation, Nanyang Technological University, Singapore.

Singapore Examinations and Assessment Board. (2008). *GCE 'A' level-H2 mathematics examination questions classified topic by topic.* Singapore: Dyna Publishers.

Singapore Examinations and Assessment Board. (2009). *PSLE Examination Questions 2004-2008: Mathematics.* Singapore: Educational Publishing House.

Thomas, D. & Brown, J. S. (2007). The play of imagination: Extending the literary mind. *Games and Culture, 2*(2), 149-172.

Part II

The Processes and Pedagogies

Chapter 2

Tasks and Pedagogies that Facilitate Mathematical Problem Solving

Peter SULLIVAN Judith MOUSLEY Robyn JORGENSEN

This is a report from one aspect of a project seeking to identify teacher actions that support mathematical problem solving. The project developed a planning and teaching model that describes the type of classroom tasks that can facilitate mathematical problem solving, the sequencing of the tasks, the nature of teaching heterogeneous groups, ways of differentiating tasks, and particular pedagogies. We report here one teacher's implementation of the model using a unit of work that he planned and taught. The report provides important insights into the implementation of the theoretically founded model and the responses of students. We found that the model can be used for planning and teaching and for encouraging problem solving. The model has a positive effect on the learning of most students. Specific teachers actions were identified in order to address the needs of the students we are most keen to support, those experiencing difficulties.

1 Introduction

In considering the nature of the curriculum and the pedagogies that are necessary to prepare students, whether in Singapore, Australia or anywhere else, for the demands of the future, for the development of society, and to ensure international competitiveness two needs must be addressed. The first is the need not only for adequate numbers of mathematics specialists operating at best international levels, capable of generating the next level of knowledge and invention, but also for mathematically expert professionals such as engineers, economists, scientists, social scientists, and planners. The second need is for the workforce to be appropriately educated in mathematics to contribute productively in an ever changing global economy, with rapid revolutions in technology and both global and local social challenges. An economy competing globally requires substantial numbers of proficient workers able to learn, adapt, create, and interpret and analyse mathematical information. Clearly it is not enough for students to become proficient in mathematical procedures, they also need to use their mathematics in unfamiliar situations and to apply knowledge from one context to other contexts. Like anything else, students can be taught to do this, which essentially needs they must have experience in creating mathematics for themselves and in solving unfamiliar problems.

It is difficult to identify unequivocal research results that can assist teachers in doing this in their everyday complex and multidimensional classrooms. We acknowledge the importance of factors such as classroom resources, organisation and climate, interpersonal interactions and relationships, social and cultural contexts, student motivation and their sense of their futures, family expectations, and organisation of schools. Nevertheless we argue that an important component of understanding teaching and improving learning is to identify the types of tasks that prompt engagement, thinking, and the making of cognitive connections, and the associated teacher actions that support the use of such tasks, including addressing the needs of individual learners. The challenge for mathematics teachers is to foster mathematical learning, and the key media for pedagogical interaction between teacher and students is the tasks in which the students engage. This is the essence of teaching problem solving.

2 Assumptions About Problem Solving and Classroom Activity

Our research is based on assumptions about posing problems and tasks, including the need for teachers to challenge all students while offering support for students experiencing difficulty. We draw on a socio-cultural perspective (Lerman, 2001) which extends the work of Vygotsky including his (1978) zone of proximal development (ZPD) which he described as the "distance between the actual developmental level as determined by independent problem solving and the level of potential development as determined by problem solving under adult guidance or in collaboration with more capable peers" (p. 86). A key aspect of the notion of the ZPD as it applies to teaching is that it defines the work of the individual or class as going beyond tasks or problems that students can solve independently, so that the students are working on challenges for which they need support. In other words, the teacher's task is to pose to the class problems that most students are not yet able to do.

Another key aspect of ZPD is that it provides a metaphor for the support that teachers can offer to students experiencing difficulty. If, for example, the teacher poses problems that are challenges for all students, in most classes there will be some students who are not already at the level of independent problem solving for this particular problem. We argue that adult guidance or peer collaboration might be offered to such students through adapting the task on which they are working, as distinct from, for example, grouping students together and having a group undertake quite different work.

3 Fostering Problem Solving by Posing Open-Ended Tasks

Within our approach, we suggest that the type of problems posed by teachers, in this case open-ended tasks, provide a way of mediating the learning between the student and mathematics. Essentially, we assume that operating on open-ended tasks can support mathematics learning by fostering operations such as investigating, creating, problematising, communicating, generalising, and coming to understand—as distinct from merely recalling—procedures.

There is a substantial support for this assumption. Examples of researchers who have found that tasks or problems that have many possible solutions contribute to such learning include those working on investigations (e.g., Wiliam, 1998), those using problem fields (e.g., Pehkonen, 1997), those exploring problem posing by students (e.g., Leung, 1997), and the open approach (e.g., Nohda & Emori, 1997). It has been shown that opening up tasks can engage students in productive exploration (Christiansen & Walther, 1986), enhance motivation through increasing the students' sense of control (Middleton, 1995), and encourage pupils to investigate, make decisions, generalise, seek patterns and connections, communicate, discuss, and identify alternatives (Sullivan, 1999). Open-ended tasks have been shown to be generally more accessible than closed examples, in that students who experience difficulty with traditional closed and abstracted questions can approach such tasks in their own ways (see Sullivan, 1999). Well-designed open-ended tasks also create opportunities for extension of mathematical operations and dimensions of thinking, since students can explore a range of options as well as considering forms of generalised response.

The tasks used as the basis of our research are an important contribution to this field in that, as well as incorporating the important positive characteristics of the above approaches they also have a specific focus on aspects of the mathematics curriculum. We describe them as *content-specific* open-ended tasks.

4 Content-Specific Open-Ended Mathematical Tasks

The nature of *content specific open-ended tasks* can best be illustrated by some examples:

> *If the perimeter of a rectangle is 24 cm, what might be the area? Draw as many different triangles as you can with an area of six square units. (Drawn on squared paper)*
>
> *The mean height of four people in this room is 155 cm. You are one of those people. Who are the other three?*
>
> *A ladder reaches 10 metres up a wall. How long might be the ladder, and what angle might it make with the wall?*
>
> *A train takes 1 minute to go past a signal. How long might the train be, and how fast might it be travelling?*
>
> *What are some functions that have a turning point at (1,2)?*
>
> *Find two objects with the same mass but different volumes.*

Such tasks are *content-specific* in that they address the type of mathematical operations that form the basis of textbooks and the conventional mathematics curriculum. Teachers can include these as part of their teaching without jeopardising students' performance on subsequent internal or external mathematics assessments.

In each open-ended task there is considerable choice in relation to operations: different strategies and solution types are possible. Some students might use trial and error to seek a variety of arithmetically derived solutions, and others may apply or develop a generalised algebraic approach using a formula and graphs, while others may satisfy themselves by exploring further combinations and perhaps discovering and employing patterns. Class discussion about the range of approaches used and range of solutions found can lead to an appreciation of their variety and relative efficiencies, key concepts like constant and variable, and the power of some mathematical methods as well as the thinking that underpins these. When all students can contribute to such discussions in their own ways, there is potential for thoughtful questioning by the teacher to draw students into new levels of engagement and learning. The tasks foster many of the aspects of problem solving.

5 Mathematical Problem Solving and our Planning and Teaching Model

We argue that teaching experiences designed to support mathematical problem solving need five key elements that can be summarised as follows.

5.1 *The tasks and their sequence*

As discussed above, open-ended tasks create opportunities for mathematical problem solving, but they also need to be effectively incorporated in a sequential development of learning. This relates closely to what Simon (1995) described as a hypothetical learning trajectory that

> ... provides the teacher with a rationale for choosing a particular instructional design; thus, I (as a teacher) make my design decisions based on my best guess of how learning might proceed. This can be seen in the thinking and planning that preceded my instructional interventions ... as well as the spontaneous decisions that I make in response to students' thinking. (pp. 135–136)

Simon (1995) noted that such a trajectory is made up of three components: the learning goal that determines the desired direction of teaching and learning, the activities to be undertaken by the teacher and students, and a hypothetical cognitive process, "a prediction of how the students' thinking and understanding will evolve in the context of the learning activities" (p. 136).

During our research, the use of sequenced open-ended tasks has improved students' engagement, as evidenced by time on task, participation in discussions, and increase in successful completion of the teaching and learning activities focusing on mathematical problems (see Sullivan, Mousley, & Zevenbergen, 2006).

5.2 *Enabling prompts*

Teachers offer enabling prompts to allow students experiencing difficulty to engage in active experiences related to the initial problem. These prompts can involve slightly lowering an aspect of the task demand, such as the form of representation, the size of the number, or the number of steps, so that a student experiencing difficult can proceed at that new level, and then if successful can proceed with the original task. This approach can be contrasted with the more common requirement that such students (a) listen to additional explanations; or (b) pursue goals substantially different from the rest of the class. The use of enabling prompts has generally resulted in students experiencing difficulties being able to start (or restart) work at their own level of understanding and enabled them to overcome barriers met at specific stages of the solving of the problems. This approach is derived from the work of Ginsburg (1997), and Griffin and Case (1997).

5.3 *Extending prompts*

Teachers pose prompts that extend the thinking of students who solve the problems readily in ways that do not make them feel that they are merely getting more of the same (see Association of Teachers of Mathematics, 1988). Students who complete the planned tasks quickly are posed supplementary tasks or questions that extend their thinking and activity. Extending prompts have proved effective in keeping higher-achieving students profitably engaged and supporting their development of higher-level, generalisable understandings.

5.4 *Explicit pedagogies*

Teachers make explicit for all students the usual practices, organisational routines, and modes of communication that impact on approaches to learning. These include ways of working and reasons for these, types of responses valued, views about legitimacy of knowledge produced, and responsibilities of individual learners. As Bernstein (1996) noted, through different methods of teaching and different backgrounds of experience, groups of students receive different messages about the overt and the hidden curriculum of schools. We have listed a range of particular strategies that teachers can use to make implicit pedagogies more explicit and so address aspects of possible disadvantage of particular groups (Sullivan et al., 2006). We have found that making expectations explicit enables a wide range of students to work purposefully, and to appreciate better the purpose of mathematical problems that are posed.

5.5 *Learning community*

A deliberate intention is that all students progress through learning experiences in ways that allow them to feel part of the class community and contribute to it, including being able to participate in reviews and summative class discussions about the work. To this end, we propose that all students will benefit from participation in at least some core problems that can form the basis of common discussions and shared experience, both social and mathematical, as well as a common basis for any following lessons and assessment items on the same topic. We have found that the use of tasks and prompts that support the participation of all students has resulted in classroom interactions that have a sense of learning community (Brown & Renshaw, 2006), with wide-ranging participation in leaning activities as well as group and whole-class discussions.

The research, reported below, is about the implementation of this teaching and planning model in a class, and this teacher's approach to teaching of subtraction using a problem solving orientation.

6 The Next Phase of the Research

The data reported below are from analysis of a sequence of lessons created and taught by one of our project teachers. For this stage of the research, we sought to

(a) examine whether teachers can use the planning and teaching model to create and teach mathematical learning experiences based on a problem solving approach;

(b) find out whether the model contributes to the goal of creating inclusive experiences; and

(c) evaluate the impact of the model and tasks on the learning of the students, especially those experiencing difficulty.

Essentially the goal of this stage of the research was to find out whether the model is feasible in classroom contexts, and to evaluate the impact of its implementation on student learning and problem solving.

While a larger number of teachers were involved in our research overall, there were five teachers who participated in all phases of the research and the associated professional development. The five teachers clearly had a strong commitment to their own professional development in that there were no incentives for their participation. In the research phase being reported in this article, each of these five teachers planned a unit of work on a topic of their choice, based on the planning and teaching model described above. They designed a pre-test, a post-test, and a sequence of activities and associated tasks to achieve their overall learning goals — usually determined by curriculum documents provided by State authorities.

Generally, each planning unit covered an extended sequence of up to eight lessons. A trained observer observed two lessons for each teacher. Her observations included both a count of specific aspects of the planning model, such as the number of *enabling prompts* posed, as well as a concurrent naturalistic summary being written. The observer or teachers collected samples of students' work. The teachers kept written records, and they were interviewed after the lessons.

The following data are from the teaching of one of these teachers, Mr Smith (not his real name), and are illustrative of the elements of the project and the teaching overall. Mr Smith was similar to the other teachers in most respects. While he was highly professional, and had an engaging personality, especially when interacting with his class, he was not chosen because of any outstanding personal or professional characteristics. Rather, he was seen to be representative of the group and how they approached their teaching. The intention for this detailed examination of one teacher's adaptation of the planning model is to offer a report on what is possible in terms of the objects of mathematical learning, the activity, the tasks and the operations; rather than, for example, considering less detailed reports of a larger number of teachers. This gives new insights into the ways students respond to this type of tasks.

We focus here on a two-week period where Mr Smith used a variety of open-ended tasks that he created. This is a representative period, and not one where the teaching and learning was outstanding in any way. Indeed the examples he created are somewhat mundane, but they do create opportunities for students to make choices about their approach and to seek patterns. We describe here the overall intent of his teaching, use extracts from the observation notes to verify incorporation of the elements of the above model, and consider the students' pre- and post-test results. By focussing on nine students who represent a range of abilities and outcomes, we seek to describe their responses to specific open-ended tasks as well as some opportunities for learning.

All project teachers considered early drafts of this report, and they verified that the report represents fairly the students' experience in their own class as well as their own experience of teaching. Mr Smith affirmed the report and the student descriptions as being accurate representations of his experience.

6.1 *Mr Smith's context and goals*

Mr Smith taught a Grade 6 class in a regional primary school, serving a community with both middle class and low SES families. The unit of work he developed focused on the topic of subtraction and was taught over two weeks for approximately one hour each day. Mr Smith gave the following as a summary of the activity:

> Further developing understanding of subtraction and the processes involved. Looking more closely at assessing students' progress with single/double digit problems (no trading), double-digit problems with trading, from 100 and from 1000 subtraction problems with trading.

In actuality, though, the tasks the children worked on included subtraction of decimals as well as whole numbers, and use of numbers above 1000, and some students added these operations spontaneously.

6.1.1 *Pre- and post-test results*

The particular focus of this chapter is on whether the open-ended approach also developed the fluency and accuracy of students at subtraction tasks. Therefore, three key questions from both the pre-test and matching post-test were selected to allow comparison of the students' skill development. The test had some open-ended items, such as "How many subtraction equations can you make using these numbers? Show examples". However, the skill development of the students can be better determined by examining responses to the following assessment items.

Question 6 consisted of 4 conventional subtraction items, set Out vertically, the easiest example being

$$\begin{array}{r} 533 \\ \underline{-\ 296} \end{array}$$

The question was scored as correct only if all four answers were correct.

Question 8 was "The Jones family completed a trip around Australia of 1389 km. When they arrived home the odometer read 40142.6 km. What might the reading have been before the trip began?

Question 10, headed "Missing Numbers", was set out like 5 ϶ 2 – ϶ ϶ 4 = 68. There was no specific prompt nor were multiple responses sought explicitly, even though these were possible.

Table 1 presents the profile of responses of students who completed both the pre-test and the post–test. The symbols √ and × are used to represent "Correct" and "Incorrect" respectively.

Table 1
Comparison of Pre- and Post-test Responses for 3 Subtraction Questions (n = 20)

	Pre × Post ×	Pre √ Post ×	Pre × Post √	Pre √ Post √
Question 6	4	2	1	13
Question 8	13	2	1	4
Question 10	10	2	3	5

From inspection, it does not appear that the two-week unit had much impact on the students' ability to complete such tasks. Most of the group were competent at skill exercises (Question 6) even at the start, and the unit did not have much impact on the students who could not complete the exercises by the end of the 2 weeks. Questions 8 and 10 were multiple step tasks requiring more than procedural fluency, and even though there were some students who could do them in the post-test but not in the pre-test, there were also some students for whom the reverse was the case.

6.1.2 *Nature of the teaching*

To illustrate the form of the teaching, the following was the first of the open-ended tasks to be described in the observer's notes: *Subtracting from 100, 1000, …* (This is termed Task A, below.) To introduce the task, Mr Smith had written the following on the board:

What might the answer be?

10	50	200	5000	10000
- ____	- ____	- ____	- ____	- ____
____	____	____	____	____

The observer recorded the beginning of the lesson as follows:

> Mr Smith directed the students to focus on the first problem on the board and to think what the answer could be. He then asked the students to write down some of the possible answers. Some clarifying questions from the students followed. In reply Mr Smith suggested that it didn't matter in what order they wrote their answers and they could use any strategy or system if they wished. Some discussion followed between a few students and Mr Smith as to the limit of whole-number answers available for the first example.

This is a clear illustration of the *explicit pedagogies* in the model above, in that Mr Smith drew the students' attention to what he considered important (use of personal strategies), to the multiplicity of possible responses, and to their role in choosing the nature of the responses, before attending to questions from the students. It is the explicit mentioning of these aspects of the students' approach to the tasks that we see as essential.

Once the class was set to work, Mr Smith then engaged individually with the students, offering encouragement and using *enabling prompts* as described in the model above. The observer recorded how he included prompts that were subtly challenging, relating to possible numbers of responses, the potential use of fractions and negative numbers, the possibility of creating a generalised system, and the use of technology. However, he continued to make his expectations of the class and of individuals explicit. The observer wrote:

> Mr Smith positively acknowledged students' queries and attempts: "Nine, well done!" [referring to the numbers of responses] "Yes, well done, there could be ten …"
> "Good question, does the answer need to be a whole number? ... No it doesn't have to be …"
> "Are you going to leave it as a decimal or a fraction?"
> He continued to assist around the room:
> "Kyal, you're looking puzzled. What could you put there? … Minus one, yes. What might the answer be? Nine…."
> Mr Smith noted John's "lovely system", and in reply to another student's query he suggested that "a system" would make the task "nice and easy to follow".
> Mr Smith kept assisting students around the room. "Alec, use my calculator. Does anyone else need a calculator?"
> Students asked Mr Smith how many examples they needed to do. "How many?" Mr Smith replied humorously, "For you fifteen, everyone else three!"
> Students were quietly engaged in this activity while Mr Smith coached students as needed.
> One student said, "I don't like carrying figures!" Mr Smith: "Sorry Buddy, if you haven't got them in, I'll say you've cheated."
> Mr Smith re-focussed a boy at the front table by coaching him, using a calculator, and reminding him to do maths first before resuming his drawing activity.

Such responses directed students' attention to elements of the task and helped to maintain their engagement, as well as proposing variations that could assist students experiencing difficulty. The task itself was graduated and so specific task variations were not necessary, but some of his comments did suggest a challenge.

The responses illustrate the conjecture that it is the task that provides the basis for the interactions between teacher and students, even those interactions that are about building personal relationships.

Mr Smith also used *extending prompts*, as illustrated in the following record by the observer:

> Mr Smith continued monitoring students' work. To one student he said, "Can you do one without zeros?"
> A query from another, "Can we have a 5 digit answer?" Discussion followed about possibilities of finishing up with a decimal or a negative number.

Mr Smith's comments, audible to the whole class, gave enough prompts to get them thinking and working along similar lines, exemplifying one way of building a learning community.

Another strategy that assisted this aim, as well as in building a sense of community, was his use of short reviews that were conducted after each phase of the lesson. These were not only teaching opportunities but also a chance to develop some common understandings that could be used as a basis for the next stage of the lesson. For example, at the end of the first phase, the observer recorded:

> Three students were then chosen to write one of their answers to the first example on the board. As a result, particular characteristics of the examples were highlighted and discussed; the need for careful spacing to denote place value, the use of a zero to assist place value separation, and the need to use the minus sign.
> The earlier discussion about having 9 or 10 possible answers was again in dispute. Students were then asked to look at the second example on the board. Students suggested that there would be "heaps and heaps" of possible answers.

As intended, all students participated in the various stages of the lesson, and all were able to contribute to each of the discussion periods as well as a significant closure activity about general principles that could be inferred from the activity. In other words, there were many instances of the class operating as a learning community.

The observer also attempted to quantify the lesson elements in each observation. In the case of this lesson, she identified 7 enabling prompts, 5 extending prompts, 2 instances of explicit pedagogies, and 5 occasions in which the teacher's intent was described as "building learning communities". In other words, this lesson, as did many of the others observed, incorporated many examples of the features of the elements of teaching proposed in the model above.

6.1.3 *Analysis of students' responses to various tasks*

To allow consideration of the impact of learning on individual students, the students' written work was later examined. Three students who were incorrect on the each of each of questions 6, 8, 10 (Jenni, John, and Eric) were identified and termed by us as the "stragglers"; 3 students who scored question 6 correct, but question 8 and 10 incorrect on both tests (Elaine, Sheryl, and Jeremy) were termed the "competent group"; and 3 students who completed all 3 questions correctly on both tests (Diane, Ellen, Becky) were termed the "achievers". The responses made by these groups of students to particular open-ended learning tasks are described in the following. The intention of this analysis was to allow detailed and comparative examination of selected students' responses to the assessments, and to the class based tasks.

Task A: Subtracting from 100, 1000,... All students gave multiple responses to the tasks, some giving more that 70 possibilities altogether. The illustrative examples presented below were given by the particular groups of students. The particular responses of the "stragglers" were as follows:

> Jenni gave more than 20 responses, most of which were simple (e.g., 10 – 1 = 9). Where she attempted difficult exercises, she got them incorrect (e.g., 200 – 199 = 111).
>
> Josh gave more than 15 responses, most simple (e.g., 5000 – 3000 = 2000), all correct.

Eric gave 6 responses, 4 were simple, and 2 were more difficult but incorrect (e.g., 50 – 21 = 28).

The responses of the "competent" group were as follows:

Elaine gave more than 20 responses: some were substantial (such as e.g., 50 – 15 = 35; 200 – 170 = 30); others were simple.

Sheryl also gave more than 20 responses. In some cases these were more complex (e.g., 200 – 64 = 136; 10000 – 9635 = 365), but the rest were simple.

Jeremy gave more than 20 responses: some simple but others more complex
(e.g., 50 – 24 = 26, 200 – 103 = 97; 10000 – 4996 = 5004).

The responses of the "achievers" were as follows:

Diane gave more than 70 responses, all correct, some decimals (e.g., 10 – 4.5 = 5.5), with many requiring exchanging before calculating a response.

Ellen gave 40 responses, all correct, with most being substantial (e.g., 10000 – 2962 = 7038).

(Becky missed this class.)

In other words, it seems that the "achievers" chose examples that extended their thinking. The open-ended nature of the task and extending prompts not only created opportunities to practise their skills, but also to extended their understanding of subtraction. The task and pedagogy also allowed the "competent" group to demonstrate competence in a range of skills and understandings, and this group used the open-ended nature of the task as well as the teacher's prompts to choose at least some examples that extended themselves. However, not all students reaped the benefits as the "stragglers" either gave responses that would not have allowed opportunity for skill practice, at least at the level of the test items, and may have even reinforced some misconceptions. The implications of this for teaching and for the model are described below.

To give a sense of some of the other lessons and tasks used by Mr Smith and the responses of these students, the following are three other open-ended tasks used as part of the unit.

Task B: Given the difference, create the question. The students were give a sheet divided into four parts, with a number in each part (respectively, 26, 982, 3193, 5.78). The students were invited to create—

and to write in that part of the paper—subtraction questions that gave that number as the answer.

Once more, all students in the class gave multiple responses, with most students giving more than 20 different possibilities. Responses of the "stragglers" for this task were:

Jenni gave more than 20 responses, most non trivial, using a pattern of responses with whole numbers mostly correctly (e.g., 3205 – 12 = 3193; 3206 – 13 = 3193), but extended the patterns to decimal numbers incorrectly (e.g., 5.79 – 1 = 5.78; 5.80 – 2 = 5.78, and so on)

Josh gave 23 responses, most trivial (e.g., 3197 – 4), and gave similar responses to Jenni for the decimal part.

Eric gave 9 responses, some non trivial (e.g., 200 – 174; 2000 – 1018). All were correct, although he did not attempt the decimal task.

Responses of the "competent" group were, once again, mixed:

Elaine gave more than 20 responses. In the first two tasks she used trading even when not necessary (e.g., 990 – 8). Her response to the third task was simple and her responses to the decimal task were incorrect like Jenni's.

Sheryl also had greater than 20 responses, generally simple, all correct with the exception of the decimals task in which the responses were also similar to Jenni's.

Jeremy gave a substantial number of correct responses to each of the tasks (e.g., 200 – 174 = 26; 1000 – 18 = 982; 4000 – 907 = 3193; 6.78 – 1.0 = 5.78).

Responses of the "achievers" again demonstrated creative solutions, the use of generalisable patterns, and extended thinking. It was clear that this group benefited once more from the open-ended challenge of the task and the teacher's extending prompts:

Diane gave 18 responses, some substantial (e.g., 333 – 307), with no errors.

Ellen gave 23 responses, many substantial (e.g., 7.94 – 2.16), with no errors.

Becky gave 15 responses, some substantial (e.g., 9.20 – 3.42), with no errors.

Of the class overall, there were 9 students who gave multiple incorrect responses, 8 students who were predominantly correct but generally used simple examples and sometimes possibly reinforced misconceptions, and 7 students whose responses that could be categorised as insightful and building on patterns (e.g., 10 – 4.22 = 5.78; 11 – 5.22 = 5.78). This suggests that the 9 focus students are fairly representative of the spread of responses overall.

It was notable that the "achievers" and the "competent" students chose examples that extended their thinking on subtraction, and at least gave them practice at the appropriate skill and conceptual level. The "stragglers" proved more likely to choose examples within their level of competence, and not beyond, and in some cases were reinforcing misconceptions. This is a key challenge for the model, and we propose a variation as is discussed below. However, all were able to participate in the whole class discussions and describe their reasoning well when asked to explain correct examples. The observer and the teacher both noted a strong sense of participation and community in this lesson, not only for the higher-achieving students.

We have noted many incidents throughout the research where relatively open-ended questions allowed teachers to see where individuals and groups of students had a misunderstanding that needed whole-class attention. With this task, for example, Jenni's misconception was common, so Mr Smith could determine where more didactic teaching would be required.

Task C: Giving an answer in a range. In this lesson, the task had two parts: "What subtraction problems would give an answer (i) between 40 and 50; and (ii) around 57?"

All students in the class gave multiple responses to the first part of this task, and most gave multiple responses to the second part. Of the "stragglers":

> Jenni gave more than 40 responses, generally non trivial. To the first part, she such gave responses such as 100 – 52, and to the second she used a pattern (e.g., 70 – 13; 71 – 14, and so on).
>
> Josh gave 5 responses to the first part, all of which were simple (e.g., 49 – 2), and none to the second task.

Eric gave 10 responses: some to the first part were substantial (e.g., 100 – 56 = 44) and likewise for some to the second task (e.g., 350 – 293 = 57).

Of the "competent" group:

Elaine gave more than 15 responses. Some responses to the first task were simple (e.g., 48 – 4 = 44), while the rest were more complex (e.g., 246 – 189 = 57).

Sheryl gave multiple responses most of which were substantial (e.g., 56 – 7 = 49; 209 – 152 = 57). All responses were correct.

Jeremy also had most responses correct, most of which were substantial (e.g., 50.2 – 4.1 = 46.1; 100 – 43 = 57).

Of the "achievers":

Diane gave more than 15 responses, most substantial (e.g., 62 – 14 = 48; 249 – 192 = 57) to the respective tasks.

Ellen gave 14 responses, all substantial (e.g., 70.29 – 28.14 = 42.15; 222 – 165 = 57).

Becky had more than 14 responses, most of which were substantial such as 235 – 185 = 50 and 626 – 569 = 57.

All students participated well throughout the lesson and their work showed evidence of attention to Mr Smith's subtle prompts and challenges. It seems that Eric (a "straggler") as well as all the "competent" students and the "achievers" were working at the level of the items in Question 6, and close to the complexity of the tasks implied by Question 10. Other than Jenni and Josh, all of these students gave substantial responses to parts of the task. However, it seemed that Jenni also did some productive work, although below the complexity of the Question 6 items. This is discussed further below.

Note that for the "stragglers" and "competent" group, the responses were generally less sophisticated than required by Questions 8 and 10 on the tests, but not by much. It would be reasonable to assume from observation alone that the open-ended classroom tasks were successful in promoting both physical and conceptual engagement throughout the lesson period, the class was progressing well. The observer noted that there was an atmosphere of communal learning with the "stragglers", in particular, participating in the lesson's review stage.

Task D: What's wrong: Simulating correction of subtraction questions. Mr Smith told the class that he had completed five subtraction exercises which he wrote on the board horizontally (e.g., 100 – 21 = 89), with some correct and some incorrect; and also five calculations presented vertically that also had also had some correct and some incorrect answers. The latter were set out like:

$$\begin{array}{r} 467 \\ -\underline{298} \\ \underline{331} \end{array}$$

Mr Smith asked the class to work out which were correct and which were not, and to advise him on how to avoid the errors in the future. In our view, this is an excellent task for both school students and student teachers in that it invites them to consider some common subtraction computational errors, and the nature of possible advice.

Mr Smith demonstrated explicit pedagogy by being specific about the task, saying that he expected the students to think of a range of possible causes for the errors. The responses of the class overall indicated that it was a successful lesson. As it happens, Jenni, Diane, Ellen were absent for this class.

In terms of the "stragglers", both Josh and Eric correctly scored the responses appropriately as either correct or incorrect respectively, but gave relatively superficial advice indicating that they had not identified any patterns of errors. Jeremy corrected the examples appropriately, and noticed the patterns in the responses, providing thoughtful advice. It would seem that the challenge of this classroom task, that Jeremy was able to respond to, was more substantial than the questions posed on the test.

For the "competent" group, Elaine provided corrected responses to the incorrect examples. In her advice she said, "Mr Smith you need to carry and you have to stop adding instead of taking, look at the signs and start concentrating, don't rush, and take your time". Sheryl also corrected her examples well, she gave correct but relatively simplistic advice that did not recognize the pattern of errors evident in the responses.

For the "achievers", Becky offered detailed and sophisticated advice, indicating that she recognised the patterns of errors and could articulate the patterns and erroneous thinking involved.

All students seemed to demonstrate competence and fluency with the calculations, and the best students and some of the competent group showed deeper insight and evidence of error analysis. In fact, given the apparent success of these learning experiences, it would have been anticipated that there would be more improvement shown on the post-test.

In summarising the performance of the students overall, both in class and on the tests, Jenni and Josh were able to participate in all of the tasks but reinforced some misconceptions at times. While doing respectable work in class and engaging at all stages of the lessons, they did not reach the standard required by question 6 of the pre- and post-tests (3 digit subtraction with trading), so it was not surprising that they did not improve on question 6 on the post-test. Eric showed improvement in class and we might have anticipated improvement in his scores, but he did not demonstrate greater skill or understanding on the test Elaine and Sheryl were clearly working at the level of Question 6, but not beyond, and we would not have anticipated that they would correctly answer 8 and 10. Jeremy did well and we could have anticipated improvement. All three "achievers" coped well with the tasks and achieved well in the tests.

6.1.4 *The delayed post test*

To examine further the possibly that growth did occur as a result of using open-ended questions and aspects of the pedagogical model, but over a longer period than the unit, the class was presented again with Questions 6, 8 and 10 about 4 months after the teaching of the unit-firstly in a test format, with only these three questions, and then one day later in a worksheet format.

The "stragglers" Josh and Eric scored all of Question 6 correct on the worksheet but not on the test, indicating some improvement. On the four parts of question 6, Josh achieved respectively on the three test administrations, 3 out of 4 correct, 2 out of 4 correct, and then 3 out of 4 correct. Eric showed further development with respective figures of 0 out of 4 correct, 2 out of 4 correct, and 3 out of 4 correct. Jenni had Questions 6 and 8 correct on both the delayed post test and the worksheet. She had earlier got 0 out of 4 correct (for question 6) in both the pre-test and post-test. Thus in Jenni's case the improvement was substantial.

In the "competent" group, Sheryl answered Question 6 correctly as previously on both the delayed post test and the worksheet, and also got Question 10 correct on the worksheet. Elaine also got Question 6 correct again on both original tests, then was correct on both test and worksheet for Question 10. Jeremy got all three questions correct in both forms. All the students improved, and Jeremy improved to the level of the "achievers" on the post-test.

The achievers demonstrated competent performance overall. Diane and Ellen got all three questions correct in both forms of delayed assessment. Becky was absent for the delayed post-test and worksheet.

The overall longer-term improvement is of interest because the teacher, Mr Smith, reported that there has been no explicit teaching of subtraction in the intervening period. Even though it is not possible to identify the impetus for the improvement in the skill levels of these students, it is possible that the nature of the experiences in the subtraction unit created sufficient awareness of the conceptual possibilities and skill development to support the potential for further growth to continue after the teaching period.

7 Summary and Conclusion

The intention of the research was to examine whether the planning and teaching model, which used a problem solving approach based on content specific open-ended tasks, is feasible and effective in classroom contexts, whether it contributed to the goal of creating inclusive experiences, and what is the nature of the learning of the students, especially those experiencing difficulty.

This report is of one teacher's implementation of the model, and the impact on the learning of his students. It should be noted that the lessons were those created by the teacher for a unit of work with a focus on subtraction as curriculum content, and while they contained some useful open-ended tasks we do not claim that they are exemplary, or even carefully sequenced. However, the lesson descriptions illustrate specific features of the model being implemented. Mr Smith's use of the model was fairly typical of that of the other project teachers, and there were similar responses from selected groups of target students in other classrooms.

In terms of the goal of creating inclusive experiences for the wide range of student capabilities that one finds in mathematics classrooms, in the lessons reported, competent students increased their mathematical proficiency and managed to progress their own learning through undertaking the subtraction tasks. The more capable students generally reacted positively to the challenge of open-ended tasks and further opportunities for problem solving were stimulated by extending prompts offered by the teacher. The more capable students created some examples that were much more difficult than those they would have faced with traditional textbook tasks, and it was clear that they remained productively engaged, both physically and cognitively, and enjoyed the challenge as well as some subtle competition and interesting discoveries.

The focus of our interest in our project overall was on students experiencing difficulties, especially those from particular equity groups. In this case, we did not seek data on socioeconomic background or other factors, but were interested in the responses of students to the mathematics tasks. It was clear that less capable students needed closer attention from the teacher but at least they were able to participate fully in the lessons, to contribute to discussions, and to use and explain strategies that were meaningful to them. They responded well to explicit instructions and were all able to commence the tasks then listen to, and possibly benefit from, enabling prompts offered by the teacher. However, it is clearly necessary for teachers using such tasks to monitor the work of students experiencing difficulties to ensure that they are extending their current levels of competence and understanding, to provide teaching or enabling prompts in order to support such students as required, and also to ensure that they are not merely reinforcing misconceptions by practising incorrect procedures. This point will be incorporated into our model of planning and teaching.

In terms of the planning and teaching model, we examined whether teachers, including the one reported on above, could use the model, incorporating the use of a sequence of open-ended tasks that could create opportunities for personal constructive activity by students; *enabling prompts* to allow those experiencing difficulty to engage in the class work; and supplementary, *extending prompts* for students who complete the initial task readily. This report indicates that it was possible for the teacher to plan and teach a unit of work based around content specific open-ended questions that engaged the students in mathematical experiences, building students' skills from their current levels, and utilising prompts as appropriate. We note that the open-ended tasks that Mr Smith developed were less challenging that the ones used by other project teachers, but we stress that these tasks were his creation, and in any case were likely to allow more opportunities for problem solving than would comparable closed text book exercises.

We also explored the nature of constraints experienced, but teachers reported none and few were obvious to observers. All of the teachers reported that the students were willing to take the necessary risks, and in this case, the better students seemed willing to take the most risk, which is contrary to the Dweck (2000) hypothesis. The "stragglers" did not extend themselves in relation to skill development. Generally, the teachers reported that they were comfortable using each of the aspects of the model in planning lessons and conducting them, although as Mr Smith reported, "It takes time to come to grips with the range of strategies".

We also recorded evidence of the learning by the students of particular skills, because it appears that teachers generally are reluctant to experiment with alternate strategies because of potential threats to the skill learning of students. We noted that merely examining differences between matching pre- and post-test items did not fully illustrate the students' mathematical development. In fact, while there was limited improvement during the course of the unit of work, there was substantial improvement after the end of the teaching of Mr Smith's unit of work — and this was also experienced in several other classrooms. It is possible that this improvement may have been a result of awareness that was created through the use of challenging open-ended tasks. This is a question worthy of further exploration. Somewhat connected is the possibility that either the original time lapse of two weeks or the test itself did not allow the measurement of growth. In other words, Question 6 may not have been complex enough to detect smaller amounts of growth by the "stragglers" and Questions 8 and 10 might have been too much of a leap for the "competent" group. The nature of classroom assessment of skill learning may also require further investigation.

References

Association of Teachers of Mathematics (1988). *Reflections on teacher intervention.* Derby, UK: ATM.

Bernstein, B. (1996). *Pedagogy, symbolic control, and identity: Theory, research, critique.* London: Taylor & Francis.

Brown, R., & Renshaw, P. (2006). Transforming practice: Using collective argumentation to bring about change in a year 7 mathematics classroom. In P. Grootenboer, R. Zevenbergen, & M. Chinnapan (Eds.), *Proceedings of the 29th Conference of the Mathematics Education Research Group of Australasia* (pp. 99–107). Canberra: MERGA.

Christiansen, B., & Walther, G. (1986). Task and activity. In B. Christiansen, A. G. Howson, & M. Otte (Eds.), *Perspectives on Mathematics Education* (pp. 243–307). Dordrecht: Reidel.

Dweck, C. S. (2000). *Self theories: Their role in motivation, personality, and development.* Philadelphia: Psychology Press.

Ginsburg, H. P. (1997). Mathematical learning disabilities: A view for developmental psychology. *Journal of Learning Disabilities, 30*(1), 20–33.

Griffin, S., & Case, R. (1997). Re-thinking the primary school with curriculum: An approach based on cognitive science. *Issues in Education, 3*(1), 1–49.

Lerman, S. (2001). Cultural, discursive psychology: A socio-cultural approach to studying the teaching and learning of mathematics. *Education Studies in Mathematics, 46*(1–3), 87–113.

Leung, S. S. (1998). On the open-ended nature of mathematical problem solving. In E. Pehkonen (Ed.), *Use of open-ended problems in mathematics classrooms* (pp. 26–35). Helsinki: Department of Teacher Education, University of Helsinki.

Middleton, J. A. (1995). A study of intrinsic motivation in the mathematics classroom: A personal construct approach. *Journal for Research in Mathematics Education, 26*(3), 254–279.

Nohda, N., & Emori, H. (1997). Communication and negotiation through open approach method. In E. Pehkonen (Ed.), *Use of open-ended problems in mathematics classrooms* (pp. 63–72). Helsinki: Department of Teacher Education, University of Helsinki.

Pehkonen, E. (1997). Use of problem fields as a method for educational change. In E. Pehkonen (Ed.), *Use of open-ended problems in mathematics classrooms* (pp. 73–84). Helsinki: Department of Teacher Education, University of Helsinki.

Simon, M. (1995). Reconstructing mathematics pedagogy from a constructivist perspective. *Journal for Research in Mathematics Education, 26*, 114–145.

Sullivan, P. (1999). Seeking a rationale for particular classroom tasks and activities. In J. M. Truran & K. N. Truran (Eds.), *Making the difference. Proceedings of the 21st annual conference of the Mathematics Educational Research Group of Australasia* (pp. 15–29). Adelaide: MERGA.

Sullivan, P., Mousley, J., & Zevenbergen, R. (2006). Developing guidelines for teachers helping students experiencing difficulty in learning mathematics. In P. Grootenboer, R. Zevenbergen & M. Chinnappan (Eds.), *Identities, cultures and learning spaces, proceedings of the 29th annual conference of the Mathematics Education Research Group of Australasia* (pp. 496–503). Sydney: MERGA.

Vygotsky, V. (1978). *Mind in society.* Cambridge, MA: Harvard University Press.

Wiliam, D. (1998). *Open beginnings and open ends.* Unpublished manuscript.

Chapter 3

Learning through Productive Failure in Mathematical Problem Solving

Manu KAPUR

Findings from two quasi-experimental studies on productive failure for a two-week curricular unit on average speed are summarized. In the first study, 75 year seven mathematics students from a Singapore school experienced either a traditional lecture and practice teaching cycle or a productive failure cycle, where they solved complex, ill-structured problems in small groups without the provision of any support or scaffolds up until a teacher-led consolidation lecture. Despite seemingly failing in their collective and individual problem-solving efforts, students from the productive failure condition significantly outperformed their counterparts from the lecture and practice condition on both the well-structured and higher-order application problems on the post-test. A second study with 109 year seven students from the same school replicated and extended these findings. Compared with students who experienced scaffolded solving of complex, ill-structured problems, students in the productive failure condition demonstrated greater representation flexibility in working with graphical representations. Findings and implications of productive failure for mathematics teaching and learning are discussed.

1 Introduction

When and how to design structure in learning and problem-solving activities is a fundamental research and design issue in education and the learning sciences. Structure can be operationalized in a variety of forms such as structuring the problem itself, scaffolding, instructional facilitation, provision of tools, content support, expert help, and so on (e.g., Hmelo-Silver, Duncan, & Chinn, 2007; Schmidt, Loyens, van Gog, & Pass, 2007). Thus conceived, structure is designed to constrain or reduce the degrees of freedom in learning and problem solving activities; the lower the degree of freedom, the greater the structure (Woods, Bruner, & Ross, 1976). By doing so, structure increases the likelihood of novices achieving performance success during problem solving, which they might not otherwise be able to in the absence of support structures. Indeed, a vast body of research supports the efficacy of such an approach. For example, when learners are provided with strong support structures in the form of worked solution examples before problem solving, it leads to better schema acquisition and learning (Sweller, 1988; Sweller & Chandler, 1991). This has led some researchers to argue that instruction should be heavily guided especially at the start, for without it, little if any learning takes place (e.g., Kirschner, Sweller, & Clark, 2006). Further support for starting with greater structure in learning and problem solving activities with a gradual reduction (or fading) over time as learners gain expertise comes from other research programs on scaffolding and fading (e.g., Hmelo-Silver, 2004; Puntambekar & Hübscher, 2005; Vygotsky, 1978; Woods et al., 1976)

More often than not therefore, both researchers and practitioners have tended to focus on ways of structuring learning and problem-solving activities so as to achieve performance success, whereas the role of failure in learning and problem solving much as it is intuitively compelling remains largely underdetermined and under-researched by comparison (Clifford, 1984; Schmidt & Bjork, 1992). What is perhaps more problematic is that an emphasis on achieving performance success has in turn led to a commonly-held belief that there is little efficacy in novices solving problems without the provision of support structures. While this belief may well be grounded in empirical evidence, it is also possible that by engaging novices to persist and even fail at tasks that are beyond their skills and abilities can be a productive exercise in failure.

2 Arguments Supporting the Case for Productive Failure

I present three arguments supporting the abovementioned possibility:

Argument from measurement. We make use of measures such as tests, interviews, and so on, to make inferences about students' learning. It is one thing to infer learning from observed success on measures of performance. But the conclusion that a lack of success on those measures implies a lack of learning does not logically follow. In other words, even if *A* (success on performance measures) were to imply *B* (learning), not-*A* does not necessarily imply not-*B*. A lack of learning is not a logical necessity that follows from a lack of performance; one is limited by the validity and scope of the measures of performance one adopts (Chatterji, 2003).

Argument from theory. It is also reasonable to argue that external support structures and scaffold may create a lock-in that restricts a fuller exploration of the problem and solution spaces (Reiser, 2004). While this lock-in may be effective in constraining the degree of freedom in a task thereby helping learners accomplish the task efficiently, such learning may not be sufficiently flexible and adaptive in the longer term, especially when learners are faced with novel problems (Schwartz & Martin, 2004). On the other hand, without such a lock-in, learners may explore, struggle, and even fail at solving problems. The process may well be less efficient in the shorter term but it may also allow for learning that is potentially more flexible and adaptive in the longer term. Persisting with such a process may engender increasingly high levels of complexity in the exploration of the problem and solution spaces. In turn, this build-up of complexity may allow for learning that is potentially more flexible and adaptive (Kauffman, 1995). Evidence from expert-novice literature strongly supports the notion that it is the complexity, density, and interconnectedness of conceptual schemas that differentiate experts from novices (Chi, Feltovich, & Glaser, 1981; Hardiman, Dufresne, & Mestre, 1989).

Argument from past research. Several scholars speak to the role of failure in learning and problem solving. Clifford (1979)'s review of theories related to the effect of failure (e.g., learner frustration, attribution and achievement motivation) led her to conclude that "educators who teach by the maxim, "Nothing succeeds like success," at least sometimes maybe doing more harm than good" (pg. 44). She further postulated that not only is performance success compatible with failure experiences but may *at times* be ensured by it (Clifford, 1984). There is also a growing body of supporting empirical evidence in educational research. For example, research on *impasse-driven learning* (Van Lehn, Siler, Murray, Yamauchi, & Baggett, 2003) and *preparation for future learning* (Schwartz & Bransford, 1998) provide strong evidence for the role of failure in learning.

My own work on productive failure examined students solving complex, ill-structured problems *without* the provision on any external support structures (Kapur, 2008; Kapur & Kinzer, 2009). I asked year eleven student triads from seven high schools in India to solve either ill- or well-structured physics problems in a synchronous, computer-supported collaborative learning (CSCL) environment. After participating in group problem solving, all students individually solved well-structured problems followed by ill-structured problems. Findings revealed that ill-structured group discussions were significantly more complex and divergent than those of their well-structured counterparts, leading to poor group performance as evidenced by the quality of solutions produced by the groups. However, findings also suggested a hidden efficacy in the complex, divergent interactional process even though it seemingly led to failure; students from groups that solved ill-structured problems outperformed their counterparts from the well-structured condition in solving the subsequent well- and ill-structured problems individually, suggesting a latent productivity in the failure. I argued that delaying the structure received by students from the ill-structured groups (who solved ill-structured problems collaboratively followed by well-structured problems individually) helped them *discern* how to structure an ill-structured problem, thereby facilitating a spontaneous transfer of problem-solving skills (Marton, 2007).

Taken together, the arguments from measurement, theory, and past research give us reason to believe that by delaying structure in the learning and problem-solving activities so as to allow learners to persist in and possibly even fail while solving complex, ill-structured problems can be a productive exercise in failure. The goal of this chapter is to describe an on-going, classroom-based research program on productive failure in mathematical problem solving at a mainstream, public school in Singapore. Two studies have been carried out thus far with successive cohorts of year seven students over the past two years. The first study, carried out with the 2007 cohort, was designed as an exploratory, proof-of-concept study of productive failure. The second study, carried out with the 2008 cohort, replicated and extended the findings of the first study. Fuller manuscripts detailing both studies have either been published or are currently under review (Kapur, in press; Kapur, under review; Kapur, Dickson, & Toh, 2008). For the purposes of this chapter, I will summarize the studies' design and procedures, and focus on the findings and their implications for mathematics teaching and learning.

3 Exploring Productive Failure in a Singapore Math Classroom

The first study was an exploratory study of productive failure targeting the curricular unit on average speed. Seventy five year seven mathematics students from two intact classes taught by the same teacher experienced either a conventional lecture and practice (LP) instructional design or a productive failure (PF) design. Both classes participated in the same number of lessons for the targeted unit totaling seven, 55-minute periods over two weeks. Thus, the amount of instructional time was held constant for the two conditions. All students took a pre- and post-test on average speed.

In the PF condition, student groups (triads) took two periods to work face-to-face on the first ill-structured problem (see Appendix A for an example). Following this, students took one period to solve two extension problems individually. The extension problems were designed as what-if scenarios that required students to consider the impact of changing one or more parameters in the group ill-structured problem. No extra support or scaffolds were provided during the group or individual problem-solving nor was any homework assigned at any stage. The PF cycle—group followed by individual problem solving—was then repeated for the next three periods using another ill-structured problem scenario and its corresponding what-if extension problems. Only during the seventh (and last) period was a consolidation lesson held where the teacher got the groups to share their problem representations and solution methods and strategies. The goal was to compare and contrast the effectiveness of those representations and solution methods. The teacher then shared the canonical ways of representing and solving the problems with the class. While doing so, the teacher explicated the concept of average speed in the context of the problems. Finally, students practiced three well-structured problems on average speed (see Appendix B for examples), and the consolidation ended with the teacher going through the solutions to these problems.

In the LP condition, students experienced teacher-led lectures guided by the course workbook. The teacher introduced a concept (e.g., average speed) to the class, worked through some examples, encouraged students to ask questions, following which students solved problems for practice. The teacher then discussed the solutions with the class. For homework, students were asked to continue with the workbook problems. Note that the worked-out examples and practice problems were typically well-structured problems with fully-specified parameters, prescriptive representations, predictive sets of solution strategies and solution paths, often leading to a single correct answer (see Appendix B for examples). This cycle of lecture, practice/homework, and feedback then repeated itself over the course of seven periods. Therefore, unlike in the PF condition, LP students did not experience a delay of structure; they received a high level of structure throughout the instructional cycle in the form of teacher-led lectures, scaffolded solving of well-structured problems, proximal feedback, and regular practice, both in-class and for homework.

3.1 *Findings*

An in-depth qualitative and quantitative analysis of group discussion transcripts and artifacts revealed that students from the productive failure condition produced a diversity of linked problem representations (e.g., iconic, graphical, proportions, algebraic, etc.) and methods—domain-general (e.g., trial and error) as well as domain-specific (e.g., proportions, algebraic manipulation)—for solving the problems but were ultimately unsuccessful in their efforts, be it in groups or individually. Solving a problem successfully means that groups were able to build on their representations to devise either domain-general and/or domain-specific strategies, develop a solution, and support it with quantitative and qualitative arguments (Anderson, 2000; Chi et al., 1981; Spiro, Feltovich, Jacobson, & Coulson, 1992).

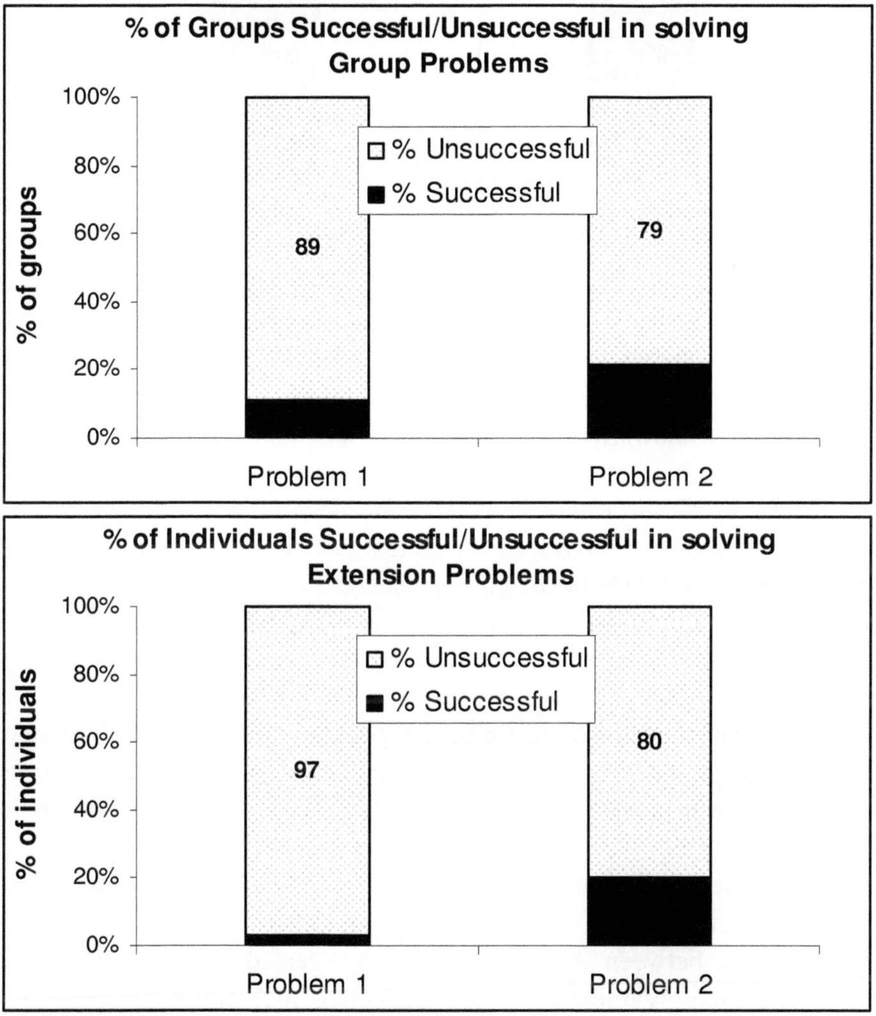

Figure 1. Success rate of groups and individual problem solving in the PF condition

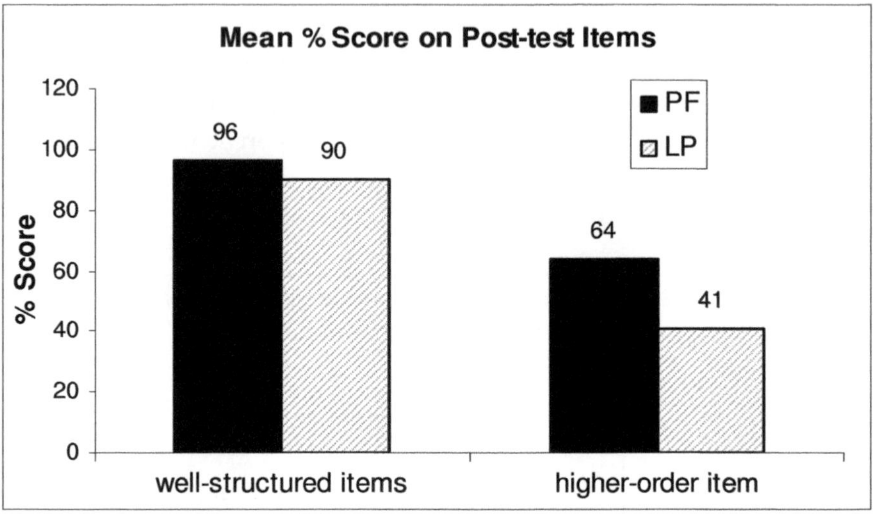

Figure 2. Performance of PF and LP students on well-structured and higher-order application items on the post-test. The y-axis represents mean score as a percentage of the maximum score on well-structured and higher-order application items.

Figure 1 shows the percentage of groups and individuals from the PF condition that were successful in solving the group and individual extension problems respectively. The success rates (indicated by the black portion of the bar graph) for groups were evidently low; only 11% (i.e., 100-89) and 21% of the groups managed to solve problems 1 and 2 respectively. Likewise, the success rates for extension problems were also low at only 3% for problem 1 and 20% for problem 2. Expectedly, students also reported low confidence in their solutions. Therefore, on conventional measures of performance success, accuracy, and efficiency, these findings may be considered a failure on the part of the PF students in spite of their persistent attempts at solving the complex, ill-structured problems.

By statistically controlling for prior knowledge as measure by students' performance on a pre-test, analysis of variance of post-test performance between the two conditions revealed (see Figure 2) that despite seemingly failing in their collective and individual problem-solving efforts, students from the PF condition significantly outperformed their counterparts from the LP condition on both well-structured and higher-order application problems on the post-tests. The difference on the well-structured items, on average, was 6%, which is remarkable given the fact that these were the very kinds of problems that LP students had routinely practiced under strong instructional scaffolds and guidance. The difference on the ill-structured item, on average, was expectedly high at 23%. In terms of effect sizes, PF students were on average .4 standard deviations above the LP students on the well-structured items, and almost one standard deviation above the LP students on the ill-structured item. This suggested that the productive failure hypothesis held up to empirical evidence even within a relatively short, two-week intervention.

However, as with any program of research, initial forays result in more questions than answers. Two major issues stood out:

a. One could always argue that perhaps students in the PF condition performed better on the post-tests because they had more collaborative activities built into the larger design. This is a perfectly valid argument that the first study was not designed to address. An immediate implication for the second study was to design the LP condition to have a similar emphasis on collaborative activities so as to unpack the effect of collaboration.

b. One could also argue that had the PF students been provided with some structure or scaffolds during their problem-solving efforts, it might have resulted in even better learning outcomes than the ones obtained in the first study (e.g., Sweller, Kirschner, & Clark, 2007). To address this issue, a third condition was added in the second study. Students in this third condition experienced all tasks and activity structures that PF students experienced except that they were provided with instructional structure and scaffolds by the teacher during group and individual problem solving.

By designing more collaborative activities in the LP condition and adding a third condition as described above, the second study built on the first study by providing stricter comparison conditions for productive failure.

4 Extending Productive Failure in a Singapore Math Classroom

The purpose of the second study was to a) replicate and extend the findings of the first study and b) unpack the effects of collaboration and instructional structure and scaffolds during activities that engage students in solving complex, ill-structured problems. One hundred and nine, year seven students from three intact classes from the 2008 cohort of the same school took part in the second study. Once again, a quasi-experimental design was used with one class assigned to the 'Productive Failure' (PF) condition, another to the 'Lecture and Practice' (LP) condition, and the third class to the 'Scaffolded Ill-structured Problem Solving' (SIPS) condition. All three classes participated in the same number of lessons for the targeted unit totaling seven, 55-minute periods over two weeks. Thus, the amount of instructional time was held constant for the three conditions. As in the first study, all students took a pre- and post-test on average speed.

The design of the PF condition was exactly the same as in the first study. The design of the LP condition was modified to incorporate a roughly equal emphasis on collaborative and individual work. The SIPS condition was designed to be exactly the same as the PF condition with one important exception. Whereas students in the PF condition did not receive any form of structure or scaffolding during the group or individual problem solving process, students in the SIPS condition were scaffolded during that process. This kind of scaffolding was typically in the form of teacher clarifications, focusing attention on significant issues or parameters in the problem, question prompts that engender student elaboration and explanations, and mini lectures and whole-class discussions to target one or a few critical aspects of problem solving (Hmelo-Silver et al., 2007; Puntambekar & Hübscher, 2005; Schmidt et al., 2007; Woods et al., 1976). After the scaffolded problem solving phase, the teacher-led consolidation lesson was the same as in the PF condition.

4.1 *Findings*

Figure 3 shows the percentage of groups from the PF and SIPS conditions that were successful in solving the group problems. The success rates (indicated by the black portion of the bar graph) for PF groups were evidently low; only 17% and 8% of the groups managed to solve problems 1 and 2 respectively. In contrast, the success rates for SIPS groups were about five times greater; 58% and 67% of the groups managed to solve problems 1 and 2 respectively. Thus, the average success rate was 62.5%, which was expectedly higher than that for the PF condition because students were given instructional scaffolding by the teacher in the form of representation scaffolds, instructional prompts and discussion, and problem-solving strategies.

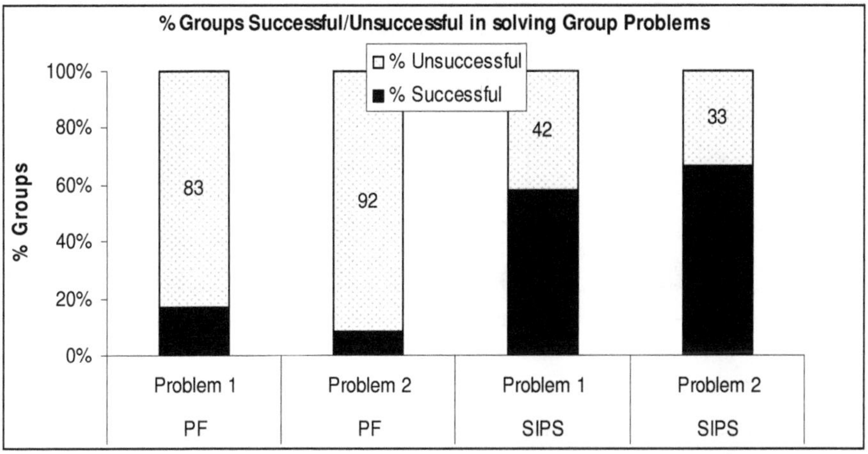

Figure 3. Success rate of group problem solving in the PF and SIPS conditions

Likewise, Figure 4 shows the percentage of students from the PF and SIPS conditions that were successful in solving the individual extension problems. The success rates for PF groups were again low; only 11% and 8% of the PF students managed to solve problems 1 and 2 respectively. In contrast, the success rates for SIPS groups were about four to seven times greater; 45% and 61% of the groups managed to solve problems 1 and 2 respectively. The higher group and individual problem solving success rates for SIPS students was not surprising, because unlike PF students, SIPS students were given instructional scaffolding by the teacher in the form of representation scaffolds, instructional prompts and discussion, and problem-solving strategies. This was further reflected in the finding that self-reported confidence in the solutions reported by SIPS students was on average twice as high as that reported by PF students—an effect size of more than 1.5 standard deviations! Thus, on conventional measures of efficiency, accuracy, and performance success, students in the PF condition seemed to have failed relative to their counterparts in the SIPS and LP conditions.

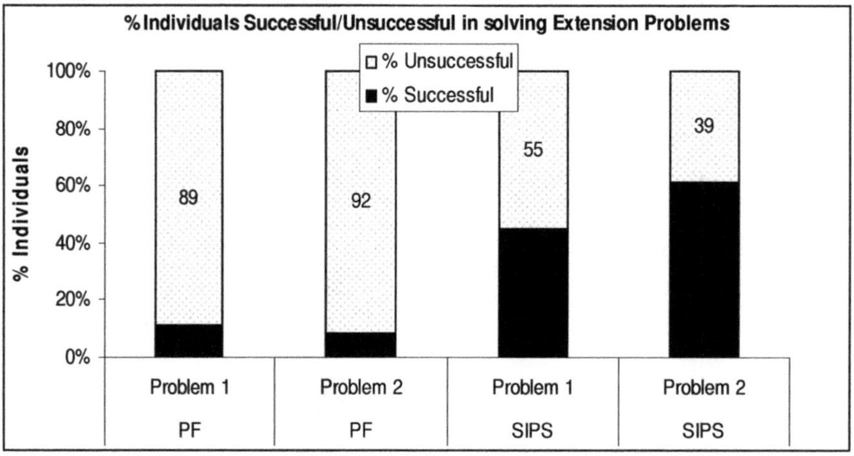

Figure 4. Success rate of individual problem solving in the PF and SIPS conditions

As in the first study, post-test performance revealed quite a different story. For the second study, the post-test comprised three well-structured problem items similar to those on the pre-test, one higher-order application item, and two additional items designed to measure *representational flexibility*, that is, the extent to which students are able to flexibly adapt their understanding of the concepts of average speed to solve problems that involve tabular and graphical representations. Note that the tabular and graphical representations were not targeted during instruction. Appendix C provides examples of the four types of items on post-test 1. Figure 5 presents the breakdown of post-test performance as a percentage of maximum score on the four types of items.

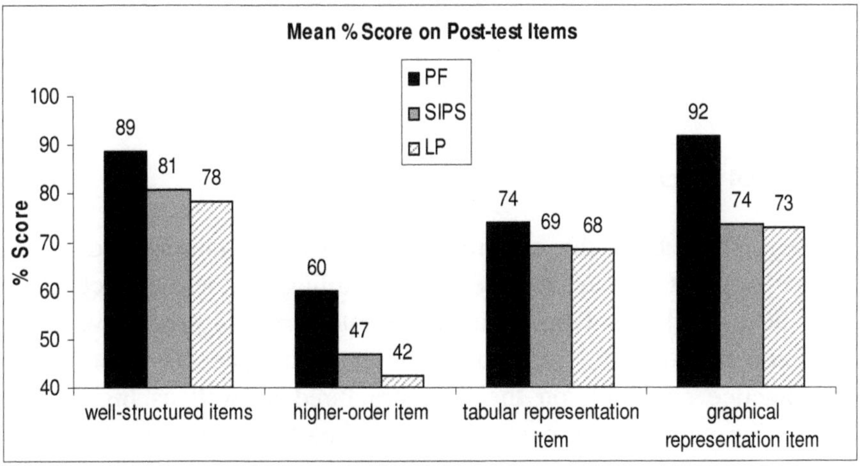

Figure 5. Breakdown of post-test 1 performance as a percentage of the maximum score for the four types of items

By statistically controlling for prior knowledge as measured by students' performance on a pre-test, analysis of variance of post-test performance between the three conditions revealed that students from the PF condition significantly outperformed their counterparts from the LP and SIPS conditions on both the well-structured items (8-11%; an effect size of 0.6 standard deviations) as well as the higher-order application item (13-18%; an effect size of 0.5 standard deviations), thereby suggesting that the productive failure hypothesis held up to empirical evidence (see Figure 2). The differences between SIPS and LP conditions were not significant, though students from the SIPS class performed marginally better than those from the LP class on both the types of items. With regard to representational flexibility as measured by performance on the tabular and graphical representation items, there were no significant differences between the conditions on the tabular representation item. This could be because of the relative concreteness of a tabular representation, which may have been easier for students to work with than a more abstract representation. However, on the graphical representation item, students from the PF condition significantly outperformed their counterparts from the SIPS and LP condition by 18-

19% % (an effect size of 1.5 standard deviations). Overall, the descriptive trend PF > SIPS > LP seemed surprisingly consistent across the different types of items.

5 General Discussion

Findings from both the classroom-based studies summarized in this chapter suggest that despite seemingly failing in their collective and individual problem-solving efforts, students from the productive failure condition significantly outperformed their counterparts from the lecture and practice condition on the well-structured as well as higher-order application items on the post-test. What is particularly interesting is the fact that students from the productive failure condition outperformed their counterparts in the other condition on the well-structured problems on the post-test, the very kinds of problems that students in the lecture and practice condition had solved repeatedly under strong instructional guidance and support. More importantly, extending the findings of the first study, findings of the second study suggest that when compared with students from the scaffolded, ill-structured problem solving condition, students from the productive failure condition performed better on both the well-structured and higher-order application items on the post-test. They also demonstrated greater representational flexibility in building upon and adapting what they had learnt to solve problems involving graphical representations—a representation that was not covered during the instructional phase.

5.1 *Explaining productive failure*

There are three interconnected explanations for productive failure.

a. *Learning to collaborate.* The first explanation deals with the notion that perhaps PF students learned how to collaborate with one another over the course of two weeks, so that they could benefit from the collaboration. However, collaboration alone cannot explain the findings because students in the SIPS and LP condition were also involved in collaboration.

b. *Learning the targeted mathematical concept.* The second explanation deals with the notion that perhaps PF students learned the mathematical concept of average speed better. Perhaps what was happening in the PF condition was that students were seeking to assemble or structure key ideas, concepts, representations, and methods while attempting to represent and solve the ill-structured problems, even though these efforts were evidently not successful in the shorter term (e.g., Amit & Fried, 2005; Chi et al., 1981; Even, 1998; Kapur, 2008). It is plausible therefore that having explored various representations and methods for solving the complex ill-structured problems prepared them to better discern and understand those very concepts, representation, and methods when presented in a well-assembled, structured form during the consolidation lesson (Marton, 2007; Schwartz & Bransford, 1998; Spiro et al., 1992). In other words, when the teacher explained the "correct" representations and methods for solving the problem, they perhaps better understood not only why the correct representations and methods work but also why the "incorrect" ones, the ones they tried, did not work (Greeno, Smith, & Moore, 1993). This very process might also explain the representational flexibility demonstrated by students from the PF condition.

c. *Developing epistemic resources for mathematical problem solving.* The third explanation deals with the notion that perhaps PF students had greater opportunities to learn how to solve mathematics problems. This notion leverages the distinction between "learning about" a discipline (as in the second explanation (b) above) and "learning to be" like a member of that discipline (Thomas & Brown, 2007). The acts of representing problems, developing domain-general and specific methods, flexibly adapting or inventing new representations and methods when others do not work, critiquing, elaborating, explaining to each other, and ultimately not giving up but persisting in solving complex problems are epistemic resources that mathematicians commonly demonstrate and leverage in their practice. Perhaps the PF design helped student expand their repertoire of epistemic resources situated within the context of classroom-based problem solving activity structures (Hammer, Elby, Scherr, & Redish, 2005). Perhaps these were the very resources they leveraged to solve the post-test problems better. To be clear: I am not arguing that there was some larger epistemological shift that took place within a two-week intervention. What I am arguing instead is that perhaps the PF design provided students with the opportunities to take the first steps towards developing these context-dependent, epistemic resources (Hammer et al., 2005). The more such opportunities are designed for students, the better they will develop such epistemic resources.

5.2 *Implications for teaching and learning design*

Many instructional designs make either implicit or explicit commitments to a *performance success* focus (Clifford, 1979, 1984; Schmidt & Bjork, 1992). A focus on achieving performance success, therefore, clearly necessitates the provision of relevant support structures and scaffolds during problem solving. In designing for productive failure, the focus was more on students persisting in problem solving than on actually being able to solve the problem successfully. In contrast to a focus on achieving performance success, a focus on *persistence* does not necessitate a provision of support structures as long as the design of the problem allows students to make some inroads into exploring the problem and solution spaces without necessarily solving the problem successfully. An important implication for the design of problems and problem solving activities is that there is efficacy in persistence itself even though it may not lead to success in performance.

However, this only begs the question: How does one design for persistence? In productive failure, designing for persistence minimally involved five interconnected principles:

a. *Designing complex, ill-structured tasks.* Two ill-structured problems were designed such that they possessed many problem parameters with varying degrees of specificity and relevance, as can be seen in the problem scenario in Appendix A. Some of the parameters interacted with each other such that their effect could not be examined in isolation. As a result, the ill-structured problem scenarios were complex, possessed multiple solution paths leading to multiple solutions (as opposed to a single correct answer), and often required students to make and justify assumptions (Jonassen, 2000; Spiro et al., 1992; Voss, 1988). In contrast, well-structured problems commonly found in textbooks afford normative representations and methods for solving them, which often resulting in a single correct answer. In such cases, either a student is able to solve the problem quickly or simply gives up. Hence, well-structured problems often do not afford opportunities for students to persist in problem solving.

b. *Designing collaborative activities.* The activity structure of collaboration helps students persist in solving problems more than what they may do individually. Hence, the choice of having students engage in collaborative problem solving was critical towards maximizing the likelihood of persistence in problem solving.

c. *Setting expectations for persistence.* It is important that teachers set appropriate expectations to assure students that it is okay not to be able to solve the ill-structured problems as long as they try various ways of solving them, especially highlighting to them the fact that there were multiple solutions to the problems. This setting of expectation is important because the usual norm in most classrooms (though not all) is not one of persistence. Instead, it is getting to the correct answer, of which there is only one, in the most efficient manner. Therefore, designing for persistence requires substantial and constant effort on the part of the teacher to set the appropriate expectations throughout the series of lessons.

d. *Withholding assistance.* It is also important for teachers to get comfortable with the idea of withholding assistance or help when students ask for it, and instead get students to try working through the problem themselves first. Students are used to asking their teachers for help so much so that they do so even before trying to figure out an answer by themselves, be it individually or in groups. At the same time, teachers are just as used to offering help and assistance when it is asked for so much so that sometimes opportunities for students to persist in solving the problem are missed; opportunities that are critical for realizing productive failure. In many ways, the first three principles of designing ill-structured problems, collaboration, and setting appropriate expectations may come to naught if teachers do not withhold assistance during initial problem solving.

e. *Iterative design*. Finally, it is important to note that designing for persistence is not a one-off design effort. Usually, one does not get it right the first time around. Decisions around the above design principles are not made in isolation but as part of an iterative design process that involves other teachers and students so that the complexity of the ill-structured problem scenarios can be developmentally calibrated with the age, grade, and ability level of the students. Before classroom implementation, multiple pilot tests with two to three groups of students are used to provide insights into and help fine-tune the design decisions described above. Classroom implementation provides additional insights that lead to further iterations and fine tuning of the design.

The abovementioned five principles are but one set of principles for designing for persistence. They are surely not the only way of doing so. Needless to say, an emphasis on persistence comes with its own set of problems because of students have varying levels of persistence, not all students persist in problem-solving, the nature of their persistence varies, and relationship between the extent to which students persist and the nature of their persistence relates to learning remains an open and important question for future research.

6 Conclusion

In the classrooms that I have been working in, the conventional bias has typically been towards heavy structuring of instructional activities right from the start. The basic argument being - why waste time letting learners make mistakes when you could give them the correct understandings? This arguably makes for an efficient process but what productive failure suggests is that processes that may seem to be inefficient and divergent in the short term potentially have a hidden efficacy about them provided one could extract that efficacy. The implication being that by not overly structuring the early learning and problem solving experiences of learners and leaving them to persist and possibly fail can be a productive exercise in failure. I contend that the work described in this chapter opens up an exciting line of inquiry into the hidden efficacies in ill-structured, problem-solving activities. Perhaps one should resist the near-default rush to structure learning and problem-solving activities for it may well be more fruitful to first investigate conditions under which instructional designs lead to productive failure as opposed to just failure.

Acknowledgements

The research reported in this paper was funded by grants from the Learning Sciences Lab of the National Institute of Education of Singapore. I would like to thank the students, teachers, the head of the department of mathematics, and the principal of the participating school for their support for this project. I am particularly indebted to Leigh Dickson who was instrumental in coordinating the logistics and data collection efforts. I am also grateful to Professors Beaumie Kim, David Hung, Kate Anderson, Katerine Bielaczyc, Liam Rourke, Michael Jacobson, Sarah Davis, and Steven Zuiker for their insightful comments and suggestions on this manuscript. This chapter summarized findings from two studies on productive failure; fuller manuscripts of the first study have already been published elsewhere (Kapur ct al., 2008; Kapur, in press), whereas the fuller manuscript of the second study is currently under review (Kapur, under review).

References

Amit, M., & Fried, M. N. (2005). Multiple representations in 8th grade algebra classrooms: Are learners really getting it? In H. L. Chick & J. L. Vincent (Eds.). *Proceedings of the 29th Conference of the International Group for the Psychology of Mathematics Education, Vol. 2* (pp. 57-64), Melbourne: PME.

Anderson, J. R. (2000). *Cognitive psychology and its implications.* New York, NY: Worth.

Chatterji, M. (2003). *Designing and using tools for educational assessment.* Boston, MA: Allyn & Bacon.

Chi, M. T. H., Feltovich, P. J., & Glaser, R. (1981). Categorization and representation of physics problems by experts and novices. *Cognitive Science, 5*, 121-152.

Clifford, M. M. (1979). Effects of failure: Alternative explanations and possible implications. *Educational Psychologist, 14*, 44-52.

Clifford, M. M. (1984). Thoughts on a theory of constructive failure. *Educational Psychologist, 19*(2), 108-120.

Even, R. (1998). Factors involved in linking representations of functions. *Journal of Mathematical Behavior, 17*(1), 105-121.

Greeno, J. G., Smith, D. R., & Moore, J. L. (1993). Transfer of situated learning. In D. K. Detterman & R. J. Sternberg (Eds.), *Transfer on trial: Intelligence, cognition, and instruction* (pp. 99-167). Norwood, NJ: Ablex.

Hammer, D., Elby, A., Scherr, R. E., & Redish, E. F. (2005). Resources, framing, and transfer. In J. P. Mestre (Ed.). *Transfer of learning from a modern multidisciplinary perspective* (pp. 89-120). Greenwich, CT: Information Age.

Hardiman, P. T., Dufresne, R., & Mestre, J. P. (1989). The relation between problem categorization and problem solving among experts and novices. *Memory and Cognition, 17*(5), 627-638.

Hmelo-Silver, C. E. (2004). Problem-based learning: What and how do students learn? *Educational Psychology Review*, 235-266.

Hmelo-Silver, C. E., Duncan, R. G., & Chinn, C. A. (2007). Scaffolding and achievement in problem-based and inquiry learning: A response to Kirschner, Sweller, and Clark (2006). *Educational Psychologist, 42*(2), 99-107.

Jonassen, D. H. (2000). Towards a design theory of problem solving. *Educational Technology, Research and Development, 48*(4), 63-85.

Kapur, M. (under review). *Designing for productive failure.* Manuscript submitted for publication.

Kapur, M. (in press). Productive failure in mathematical problem solving. *Instructional Science.*

Kapur, M. (2008). Productive failure. *Cognition and Instruction, 26*(3), 379-424.

Kapur, M., Dickson, L., & Toh, P. Y. (2008). Productive failure in mathematical problem solving. In B. C. Love, K. McRae, & V. M. Sloutsky (Eds.), *Proceedings of the 30th Annual Conference of the Cognitive Science Society* (pp. 1717-1722). Austin, TX: Cognitive Science Society.

Kapur, M., & Kinzer, C. (2009). Productive failure in CSCL groups. *International Journal of Computer-Supported Collaborative Learning (ijCSCL), 4*(1), 21-46.

Kauffman, S. (1995). *At home in the universe.* New York: Oxford University Press.

Kirschner, P. A., Sweller, J., & Clark, R. E. (2006). Why minimal guidance during instruction does not work. *Educational Psychologist, 41*(2), 75-86.

Marton, F. (2007). Sameness and difference in transfer. *The Journal of the Learning Sciences, 15*(4), 499-535.

Puntambekar, S., & Hübscher, R. (2005). Tools for scaffolding students in a complex learning environment: What have we gained and what have we missed? *Educational Psychologist, 40*(1), 1-12.

Reiser, B. J. (2004). Scaffolding complex learning: The mechanisms of structuring and problematizing student work. *The Journal of the Learning Sciences, 13*(3), 423-451.

Schmidt, R. A., & Bjork, R. A. (1992). New conceptualizations of practice: Common principles in three paradigms suggest new concepts for training. *Psychological Science, 3*(4), 207-217.

Schmidt, H. G., Loyens, S. M. M., van Gog, T., & Paas, F. (2007). Problem-based learning *is* compatible with human cognitive architecture: Commentary on Kirschner, Sweller, and Clark (2006). *Educational Psychologist, 42*(2), 91-97.

Schwartz, D. L., & Bransford, J. D. (1998). A time for telling. *Cognition and Instruction, 16*(4), 475-522.

Schwartz, D. L., & Martin, T. (2004). Inventing to prepare for future learning: The hidden efficiency of encouraging original student production in statistics instruction. *Cognition and Instruction, 22*(2), 129-184.

Spiro, R. J., Feltovich, R. P., Jacobson, M. J., & Coulson, R. L. (1992). Cognitive flexibility, constructivism, and hypertext. In T. M. Duffy, & D. H. Jonassen (Eds.), *Constructivism and the technology of instruction: A conversation* (pp. 57-76). NJ: Erlbaum.

Sweller, J. (1988). Cognitive load during problem solving: Effects on learning. *Cognitive Science, 12*(2), 257-285.

Sweller, J., & Chandler, C. (1991). Evidence for cognitive load theory. *Cognition and Instruction, 8*(4), 351-362.

Sweller, J., Kirschner, P. A., & Clark, R. E. (2007). Why minimally guided teaching techniques do not work: A reply to commentaries. *Educational Psychologist, 42*(2), 115-121.

Thomas, D., & Brown, J. S. (2007). The play of imagination: Extending the literary mind. *Games and Culture, 2*(2), 149-172.

Van Lehn, K., Siler, S., Murray, C., Yamauchi, T., & Baggett, W. B. (2003). Why do only some events cause learning during human tutoring? *Cognition and Instruction, 21*(3), 209-249.

Voss, J. F. (1988). Problem solving and reasoning in ill-structured domains. In C. Antaki (Ed.), *Analyzing everyday explanation: A casebook of methods* (pp. 74-93). London: Sage Publications.

Vygotsky, L. S. (1978). *Mind in society*. Cambridge, MA: Harvard University Press.

Wood, D., Bruner, J. S., & Ross, G. (1976). The role of tutoring in problem solving. *Journal of Child Psychology and Psychiatry and Allied Disciplines, 17*, 89-100.

Appendix A

An Ill-structured Problem Scenario

It was a bright, sunny morning and the day of the Singapore Idol auditions. Hady and Jasmine were going to audition as a team. They were practicing at their friend Ken's house and were planning to bike to the auditions at Singapore Expo. The auditions were supposed to start at 2 pm and Hady and Jasmine wanted to make sure that they could make it in time.

Hady: Ken, how do we get to the Singapore Expo from here?

Ken: Well, follow this road (pointing to a map) until you reach the expressway. I usually drive at a uniform speed of 90 km/h on the expressway for about 3 minutes. After that there is a sign telling you how to get to Singapore Expo.

Jasmine: How long does it take you to reach Singapore Expo?

Ken: It normally takes me 7 minutes to drive from my house when I am traveling at an average speed of 75 km/h.

After getting the directions, Hady and Jasmine left Ken's house and biked together at Jasmine's average speed of 0.15 km/min. After biking for 25 minutes, Jasmine biked over a piece of glass and her tire went flat.

Jasmine: Oops! My tire is flat! What shall we do now? Can I just ride with you on your bike or shall we take a bus the rest of the way?

Hady: I don't think that is a good idea. My bike is old and rusty and it cannot hold both of us. Taking the bus is not a very good idea either. There is no direct bus from here to Singapore Expo, so we would have to take one bus and then transfer to another one. All the waiting for buses would definitely make us late. Do you have any money on you?

Jasmine: Let me check..... I forgot to withdraw money today. I only have $2.

Hady: I did not bring my wallet. I only have $1 for a drink.

Jasmine: Since we do not have enough money to take a taxi, shall we just leave our bikes here and walk?

Hady: It takes me approximately 5 minutes to walk to school which is about 250 meters from my home. How long does it take you to walk to school?

Jasmine: It takes me about 13-15 minutes to walk to school which is about 450 meters from my home.

Hady: No, no, no! Walking would take too much time. We will end up late. Why don't you lock up your bike and take my bike and bike ahead. Leave my bike somewhere along the route and begin walking to the audition. I will walk from here until I get to my bike and ride it the rest of the way since I can bike at a faster speed. My average biking speed is 0.2 km/min.

Jasmine: That sounds like a good idea! But how far should I ride your bike before leaving it for you and walking the rest of the way. Since we are auditioning together as a team, we have to reach there at the same time!?

How far should Jasmine ride Hady's bike so they both arrive at the audition at the same time?

Appendix B

Examples of Well-structured Problems

1. The flight distance between Singapore and Sydney is about 6316 km. A plane takes 7 h 20 min to fly from Singapore to Sydney.

a) Find the average speed of the plane from Singapore to Sydney. Give your answer correct to the nearest km/hr.

b) Sydney's time is 3 hours ahead of Singapore's time. If the plane departs from Singapore at 0955 hours, find its time of arrival in Sydney.

2. Jack walks at an average speed of 4 km/hr for one hour. He then cycles 6 km at 12 km/hr. Find his average speed for the whole journey.

Appendix C

Items on Post-test 1

A post-test well-structured item: David travels at an average speed of 4km/hr for 1 hour. He then cycles 6km at an average speed of 12 km/hr. Calculate his average speed for the entire journey.

The higher-order application item: Hummingbirds are small birds that are known for their ability to hover in mid-air by rapidly flapping their wings. Each year they migrate approximately 8583 km from Canada to Chile and then back again. The Giant Hummingbird is the largest member of the hummingbird family, weighing 18-20 gm. It measures 23cm long and it flaps its wings between 8-10 times per second. For every 18 hours of flying it requires 6 hours of rest. The Broad Tailed Hummingbird beats its wings 18 times per second. It is approximately 10-11 cm and weighs approximately 3.4 gm. For every 12 hours of flying it requires 12 hours of rest. If both birds can travel 1 km for every 550 wing flaps and they leave Canada at approximately the same time, which hummingbird will get to Chile first?

Tabular representation item: The property market has been on the rise for the past few years. In the newspaper, you find the following table with the growth rate over the past 5 years.

Year	% Growth
2003	2%
2004	7%
2005	11%
2006	14%
2007	16%

Some people are saying that the property market is growing. Other are saying that it is slowing down. Based on the table above, what do you think—is the property market growing or slowing down? Explain your answer with calculations.

Graphical representation item: Bob drove 140 miles in 2 hours and then drove 150 miles in the next 3 hours. Study the two speed-time graphs A and B carefully. Which graph - A, B, or both - can represent Bob's journey? Show your working and explain your answer.

Graph A

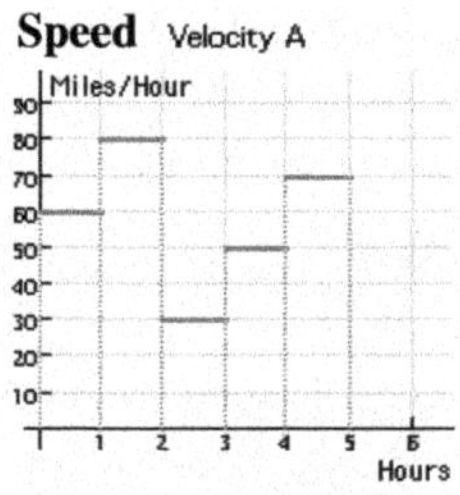

Graph B

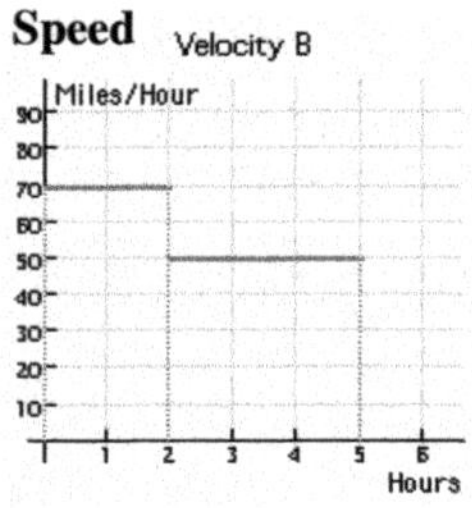

Chapter 4

Note Taking as Deliberate Pedagogy: Scaffolding Problem Solving Learning

Lillie R. ALBERT Christopher BOWEN Jessica TANSEY

This chapter provides two teaching episodes to illustrate what note taking as a tool for thought might look like in scaffolding problem solving learning. A theoretical discussion examines how note taking can be deliberate in nature, highlighting the work of Bruner and Vygotsky. This discussion includes the notion that learning and thinking depend upon internal speech, which can be developed and maintained through interaction within dynamic social contexts. The teaching episodes illustrate practical models for instructing students on how to take notes. The aim is to explore how conceptual hard scaffolding influences the interactions between teachers and their students. In addition, it includes a discussion of the role the teacher educator in scaffolding the performance of the teachers. Finally, conclusions are made regarding why the model works, noting some of the issues that surface when note taking as a tool for thinking and learning is applied in mathematics classrooms.

1 Introduction

Assigning a mathematics notebook is a common practice in middle schools (Years 5 to 8) in the United States (US). The notebook is typically a binder divided into sections consisting of work completed by students: homework, daily class-work, group-work products, quizzes, and journaling. If the notebook contains notes, they generally are incorporated into the daily class-work section and are limited to definitions and problems modeled by the teachers. Too often, this general reason for the notebook seems to be for organizational purposes. Furthermore, teachers seldom teach note-taking skills or require students to use their notes as an aid in the problem solving process. For this reason, there is very little evidence of students' use of notes beyond organizational purposes (Boch and Piolat, 2005). While these organizational purposes may validate requiring students to take notes in class, this practice does not expose students to the metacognitive aspect of note taking—writing to learn new content (Piolat, Olive, and Kellogg, 2005), nor does it provide equitable mathematical learning experiences for all types of students.

Note taking can serve as a tool for scaffolding learning. Scaffolds are the supports provided by more knowledgeable others to help a learner move from a current level of performance to a more advanced level. Essential to scaffolding within instruction is the use of language for mediation (Albert, 2000; Wertsch, 1979, 1980; Vygotsky, 1986). Note taking, as written language, is an important communicative tool that can serve as a cognitive function assisting learners in acquiring new knowledge; it can also be a tool to express thoughts. In the ethos of reform-oriented curricula, such as the *Connected Mathematics Project* (CMP) in the US, teachers need to provide students with experiences to learn how to skillfully write notes in ways that will help them develop their own knowledge and thinking about mathematics. This position is similar to a finding from a study by Boaler (2002) about the potential of reform-oriented curricula to promote equity. Boaler asserts that practices, which help students access reform approaches, may assist students in understanding "questions posed to them, teaching them to appreciate the need for written communication and justification, and discussing with them ways of interpreting contextualized questions" (p. 253).

With the publication of *Principles and Standards for School Mathematics* in 2000 (National Council of Teachers of Mathematics, 2000), the National Council of Teachers of Mathematics (NCTM) challenges mathematics teachers to reject the notion that only some students can succeed in mathematics, and to replace this commonly held belief with a philosophy that promotes equitable mathematics learning for all students. The NCTM asserts that expectations must be raised –

"mathematics can and must be learned by *all* students," (p. 13) and deliberately organized note taking may act as a tool through which mathematics teachers can promote equity in their mathematics classroom and make mathematics content more accessible to their students with diverse learning needs. To do so, teachers must deliberately increase the expectations for note taking that they hold for all students in a way that parallels the increased expectations NCTM places on them to promote equity in their classrooms. These increased expectations, together with appropriately scaffolded instruction on deliberate note taking, may lead to increased equity on performance in mathematics classrooms.

In this chapter, we provide two teaching episodes to illustrate what note taking as a tool for thought might look like in mathematics classrooms. We begin with a brief discussion of how note taking can be deliberate in nature, grounding this scheme of note taking in Bruner's (1963, 1966) idea that learning new knowledge should be a deliberate process informed by teachers' professional development. This discussion includes the Vygotskian notion that learning and thinking depend upon internal speech, which can be developed and maintained through interaction within dynamic social contexts. Next, we present teaching episodes to illustrate practical models for teaching students how to take notes. The aim is to explore aspects of the teaching episodes and how conceptual hard scaffolding influences the interactions between teachers and their students. This section also includes a discussion of the role the teacher educator play in scaffolding the performance of the teachers. Finally, we consider why the model works, noting some of the issues that surface when note taking as a tool for thinking and learning is applied in mathematics classrooms.

2 The Deliberate Nature of Note Taking: Bruner and Vygostsky

From a philosophical viewpoint, note taking in this chapter is discussed under the principle of epistemology, which encompasses the study of the origin, nature, limits, and methods of knowledge. How might teachers encourage students to use their notes to discern what they know about the mathematical knowledge learned through their problem solving reform-oriented textbooks? How does the deliberate nature of note taking scaffold students in their learning of new mathematical concepts presented in these texts? Implicated in the epistemology of mathematical learning via note taking is the work and research of Bruner (1996, 1966, & 1963) and Vygotsky (1994, 1986, 1981, & 1978). Their work is useful in explaining what processes trigger student performance when participating in practices directed by deliberate pedagogy and when identifying what resources or experiences may form the basis for student learning of mathematics. Both Bruner and Vygotsky position their notion of learning and development within a practical view in which the authenticity of learning is deduced, negotiated, and consensual. Such a view suggests that mathematical learning calls attention to the ways in which learning is changed and continuously renewed as learners interact in contexts that scaffold sense-making of the content (Driscoll, 1994). Note taking then, as a tool, allows students to demonstrate their conceptual understanding of mathematics on an abstract level in that writing is a concrete representation of thought (Albert, 2000).

Bruner (1973) asserts that learners move from a concrete understanding to an abstract understanding of the mathematical concepts they encounter. When used deliberately and purposefully, note taking acts as a bridge that connects the concrete domain to the abstract domain presented in reform-oriented texts. This can be seen when mathematics teachers apply deliberate pedagogical methods for taking notes in their classrooms; metacognition fosters students' awareness of what they need to learn, when and how they need to learn it, and self-knowledge of personal and intellectual qualities. Thus, "[k] nowing is a process, not a product" (Bruner, 1973, p. 72). From Bruner's perspective, students can best operate at a high metacognitive level when pedagogical processes are deliberate and intentional.

Like Bruner, Vygotsky (1994, 1978) believed that learning in the classroom must coalesce with deliberate pedagogy. Vygotsky introduced the notion of the *zone of proximal development* to explain how students make the transition from interpsychological functioning to intrapsychological functioning; deliberate pedagogical practices that include note taking may assist students in making this transition. The zone of proximal development is the distance between a student's "actual developmental level as determined by independent problem solving" and the higher level of "potential development as determined through problem solving under adult guidance or in collaboration with more capable peers" (1978, p. 86). This idea is useful in explaining, at least in part, why the phenomenon of using note taking as a tool for scaffolding mathematical learning makes sense. Greenfield (1984) suggests that the metaphorical nature of a scaffold, as it is known in building construction, has five attributes. "It provides a support; it functions as a tool; it extends the range of the workers; it allows the worker to accomplish a task not otherwise possible; and it is used selectively to aid the worker where needed…a scaffold would not be used for example, when a carpenter is working five feet from the ground" (p. 118). The attributes of a scaffold also make clear how note taking can be a tool that assists students in solving problems that may be difficult or unfamiliar. For example, as students apply strategies and techniques to solve difficult or unfamiliar problems, they use their written notes to help them begin to connect their thinking to mathematical ideas. Note taking provides students with opportunities to learn through writing while extending their understanding of concepts and content; the inner dialogue with *self* unequivocally offers students opportunities to write, practice, and make their thinking visual and concrete (Albert, 2000).

Note taking can be viewed as a "conceptual hard scaffold" that guides students in the problem solving process (Saye and Brush, 2002). Hard scaffolds are fixed supports or guides based on teachers' prior expectations and knowledge of difficulties students might encounter as they engage in problem solving tasks. Therefore, conceptual hard scaffolds can provide directions that help students seek relevant information to use when problem solving. Conceptual hard scaffolds may be models of approaches or processes (Simons and Klein, 2007). For example, a teacher may provide a model of a simpler problem so students can apply it to solve a complex problem. Then, the students can use their written notes as a model of how to start a difficult problem or as a hint and support in the problem solving process. Such an approach may promote higher-order thinking as well as a way for students to make connections between simple and complex problems.

Students need proper scaffolding to be introduced to the deliberate process of note taking; similarly, their teachers need proper scaffolding to learn how to teach this new approach to note taking. Professional development provides the opportunity for teachers to develop an understanding of and to improve upon the pedagogical practices needed to effectively practice deliberate note taking. The implementation of this approach to note taking is successful in the practical teaching episodes presented largely because the teachers received instruction and feedback. Their learning, just as their students' learning, is scaffolded so as to prompt a progression of cognitive functioning that results in an improved, more equitable approach to mathematics instruction.

3 Practical Applications of Deliberate Note Taking

The preceding theoretical discussion establishes the deliberate practice of note taking as a process that promotes learning, which compels students to approach mathematical knowledge and ideas with higher cognitive functions such as analysis and synthesis. It helps develop students' thinking, requiring them to evaluate their thoughts so as to organize information in ways that may not have been immediately visible to them, which in turn leads to independent thinking and problem solving. The following practical cases emerge from our research and work with two middle school teachers, Mr. Orland and Miss Lipan, using the CMP curriculum in their Algebra I classrooms. These classrooms contain students with very diverse learning styles and needs, including students who perform at an advanced level and others who have documented learning disabilities. The two teachers vary in their experiences and backgrounds; Mr. Orland is a second-year teacher and Miss Lipan is a fifth-year teacher. Both teachers originally required students to maintain notebooks and encouraged students to take notes during class discussions and interactions. The teachers seldom mentioned students' notes, however, and rarely encouraged students to use their notes beyond preparing for a weekly or unit test. As part of their professional development plan, which included goals to improve their implementation of CMP, the teachers wanted to encourage writing to learn and build collaborative groupwork into their pedagogical practices. In these collaborative groups, Mr. Orland and Miss Lipan launch multi-step problems and require written statements explaining strategies and procedures applied to solve the problems explored by students.

In Mr. Orland and Miss Lipan's previous attempts to use writing techniques for multi-step problems, they found that their students experienced difficulties in providing coherent written explanations of how they solved problems. Furthermore, Mr. Orland and Miss Lipan stated that modeling similar problems in front of the class did not seem to resolve this issue because many of the students only recorded computational information and formulas in their notebooks. Because these notes were too brief, the students did not refer to them when transitioning into group or individual work. To address this shortcoming, we developed a process that would scaffold students' learning and exploration of the studied content. The idea was to develop and implement an explicit and deliberate pedagogical model that not only engaged students in the learning process during the launching and modeling phases of the lesson but also helped them when they worked in groups (especially in the area of providing written explanations). The teacher educator, lead author of this chapter, met with Mr. Orland and Miss Lipan the day before the lesson to assist them in creating reflection questions for their students. These same questions would aid the teacher educator in focusing her observations of the classroom teachers during the launching and modeling phase of their lessons. The next day involved observing Mr. Orland and Miss Lipan model problems to their classes. The day following the lesson, the teacher educator met again with Mr. Orland and Miss Lipan to share and discuss observations and thoughts with them.

3.1 *Scaffolding in the zone of proximal development*

The practice that emerged is grounded in the Vygotskian construct of the zone of proximal development on two levels. First, the more knowledgeable other (in this case the teacher educator) works with Mr. Orland and Miss Lipan, scaffolding their learning and understanding of how to model problems for their students. Second, the Mr. Orland and Miss Lipan provided the reflection questions to students, while modeling the problems for their students. Using a think-aloud-protocol, they showed thinking and reflection in action and demonstrated to students what information is essential to include in their notes.

Mr. Orland and Miss Lipan stimulate students within their zone of proximal development through the think-aloud-protocol, which teachers use to show how they are thinking throughout the problem-solving process. The core of the zone of proximal development is the collaborative discourse between students and teachers—a social system that is actively constructed, supported, and scaffolded by the students' interactions with their teachers. The practice is designed so all students will *write to focus, write to reflect*, and *write to apply* what they are learning. When students are *writing to focus*, they are gathering relevant information about the content. In other words, the students engage in mental interaction with the teacher. At this time, students attempt to summarize what the teacher says, models, or demonstrates. Then, students *write to reflect* on key questions about the problem. At this stage, students make judgments about the content they are learning. The teacher must form the questions in a way that requires students to use their notes to grapple with the content and to extend their thinking beyond simple rote memory tasks such as recalling information or performing computational procedures. The model needs to support student learning and must equally support the pedagogy by complementing the instruction that students receive. Next, students work in groups, pairs, or independently of the teacher and *write to apply* their knowledge while referring to their notes. During this phase, students might work closely with other students as the teacher continues to scaffold their thinking. Overtime, during the academic year, students need their teacher's scaffolding less because they have repeatedly participated in the collaborative activity and made use of their written notes, and can increasingly use their notes as scaffolds. The following two episodes reflect this approach to mathematics instruction of effective note taking to solve problems.

3.2 *Episode one: Mr. Orland*

Mr. Orland launches a problem in which the objective is to model for students how to identify and compare the rate of growth in exponential relationships. He scaffolds students' progress in each mathematical learning activity, presenting the *Note Taking Model* illustrated in Figure 1. The page is folded to hide the questions so that students would focus on summarizing information presented during the launching and modeling phase of the lesson. Students made notes of the teacher modeling on the following problem (Lappan, Fey, Fitzgerald, Friel and Philips, 2004, p. 7):

> One day in the ancient kingdom of Montarek, a peasant saved the life of the king's daughter. The king was so grateful that he told the peasant she could have any reward she desired. The peasant–who was also the kingdom's chess champion–made an unusual request: "I would like you to place 1 ruba on the first square of a chessboard, 2 rubas on the second square, 4 on the third square, 8 on the fourth square, and so on, until you have covered all 64 squares. Each square will have twice as many as the previous square." When the king told the queen about the reward he had promised the peasant, the queen said, "You have promised her more money than we have in the entire royal treasury! You must convince her to accept a different reward." [The king revised the plan.] He would place 1 ruba on the first square, 3 on the next, 9 on the next, and so on. Each square would have three times as many rubas as the previous square.
>
> Make a table showing the number of rubas the king will place on squares 1 through 16 of the chessboard. As the number of squares increase, how does the number of rubas change? What does the pattern of change tell you about the peasant's reward? What is the growth factor or rate? Write an equation for the relationship between the number of the square, n, and the number of rubas, r.

Notes	*Reflection Questions*
	1. List strategies your teacher used to help him understand the problem. 2. How did the strategies help the teacher understand the problem? 3. Name another strategy you might use to solve a similar problem.

Figure 1. Note Taking Model (Adapted from Paul, 1974)

Mr. Orland presented the objective by writing it on an overhead transparency and then proceeded by analyzing the situation and showing students how to approach the problem, progressing gradually from one phase of the problem to the next. He constructed a table to show the results of the two plans, which included a discussion of how the patterns of change in the number of rubas under the two plans are similar and different. Then he wrote an equation for the relationship and used the data generated to decide which plan the peasant should take. In launching the problem, he focused on the language of the problem (e.g. underlining key terms and grappling with word meaning) and used algebraic thinking. This component required translation from a verbal representation to a symbolic representation using a letter as a variable to represent any number with the underlying aim of arriving at an expression. He articulated his thoughts orally, illustrating that the algebraic expression 2^{n-1} justifies the result that one obtains empirically by trying several particular numbers. Figure 2 contains a selection from the teacher educator's observation of Mr. Orland's launching and modeling phase of the lesson.

Teacher Educator's Observation Notes
Mr. Orland explicitly explains what he is doing to solve the problem: "I can use this picture to help me understand the problem…I am going to keep going by making a table." He illustrates metacognitive thinking: "Have I seen this problem before? How is it connected to what I have done before? I need to read the next part of the problem… I think I will look back to help me understand the problem." He reviews and reads the next part of the problem, underlining key terms (e.g., pattern of change) and highlighting unfamiliar terms (e.g., growth factor). Then he models how to look up unfamiliar terms: "I'm not sure what growth factor means; so I need to look it up…Sometimes, I forget what a term means." Mr. Orland constructs a table:

Square	Number of Rubas	
	Plan 1	Plan 2
1	1	1
2	2	3
. . . 8	. . . 128	. . . 2187

Figure 2. Observation Excerpts, Mr Orland's Lesson

After completing the modeling phase of the lesson, Mr. Orland gives students five minutes to use their notes to answer the reflection questions and then to share their answers with their group. Then, Mr. Orland engaged the whole class in a discussion about some of the strategies he used to understand and solve the problem. He asked students, "What strategies did I use to help me understand the problem?" Students responded:

S_1: "Tried one step at a time."
S_2: "Circled words you did not know."
S_3: "You looked at the picture."
S_4: "You asked yourself questions."
S_5: "You eliminated unimportant facts to get to the question…"

Next, using their notes, students completed several similar problems in which they made graphs that they compared to the graph modeled for Plan 1. Some of the students struggled with representing a general statement and using the statement to justify numerical arguments. Observation of students suggests that even when they are successfully taught symbolic manipulation, they may be unsuccessful in seeing the power of algebra as a tool for representing the general structure of a situation. For example, some students wrote $2^n - 1$ or $2n - 1$. Students needed continued scaffolding and interaction with the teacher to move toward an appropriate generalization.

3.3 *Episode two: Miss Lipan*

Miss Lipan began the modeling process by selecting and identifying for students an irregular shape, the area of which she would compute. Through a systematic approach, which she identified by name ("*the surround-and-subtract strategy*"), she promoted the development of students' understanding of how to find the area of an irregular shape. She then recorded the steps, while communicating her thinking and reasoning to students, of how to apply the *surround-and-subtract strategy*. She concluded the modeling process by reviewing the steps for using the strategy and by providing an explanation of the mathematical procedures she applied. Students wrote notes during the modeling phase of the lesson and used their notes to answer the following reflection questions: What was the name of the strategy that Miss Lipan used to solve the problem? Describe in words what she did to find the area of the shape. Name another strategy you might use to find the area of an irregular shape. Figure 3 presents a selection from the teacher educator's observation about Miss Lipan's lesson as it progressed.

Miss Lipan gave students a chance to use their notes to answer the *Guiding Questions*. After the teacher reviewed students' answers, they moved into their groups and computed the area for a number of irregular figures as well as wrote explanations that describe the strategies used to find the areas. Figure 4 shows a student's approaches to the problems. In the first problem, the student counted squares to determine the area but

Teacher Educator's Observation Notes
Miss Lipan presents the objective, which is to help students learn about how to find the area of an irregular shape. She proceeds by analyzing the situation and showing students how to approach area. "I am modeling one strategy. You need to write down what I do. I am modeling a strategy for finding the area of an irregular shape. I am going to name my strategy and prove to you how I can use it to find the area."
She draws the following, showing how to surround the original shape. In her modeling, she demonstrates how to figure out the area of a given region and, at the same time, relates it to other figures with which students are familiar, such as finding the area of a square or triangle. 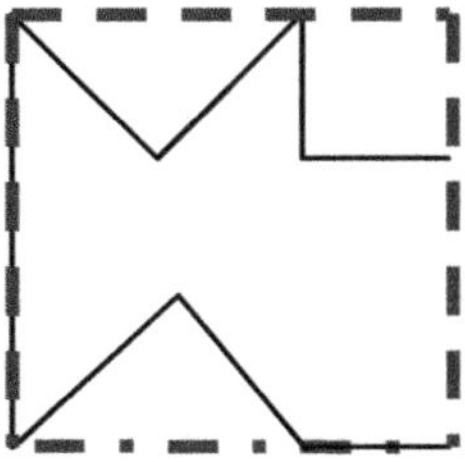
Miss Lipan records the steps for using the *surround-and-subtract strategy*, modeling how to apply the steps. "So, the total area is 9, which is 3 x 3." She provides a summary of how to apply the steps of the *surround and subtract strategy* by reviewing the steps. "First, I… then I…so I subtract from the original shape…" At this point, a student asks a question, "Does this strategy work for every irregular shape?"

Figure 3. Observation Excerpts, Miss Lipan's Lesson

applied a different approach to the next problem. For the second problem, the student explained that she divided the figure into two triangles, found the area of the triangle, and then added the area of both triangles to solve for the area of the figure. It is important to note that although some students made use of the strategy demonstrated by the teacher, many of them applied other strategies that emerged from prior experiences involving area. Then again, it was the Miss Lipan's launching and modeling of the problem that prompted students to ask themselves, "Why is this not working?" "Should I try something else?" These questions, the consistent reference to their notes, and students' continued interaction with the teacher scaffolded their application of the *surround-and-subtract strategy* and others strategies, as well.

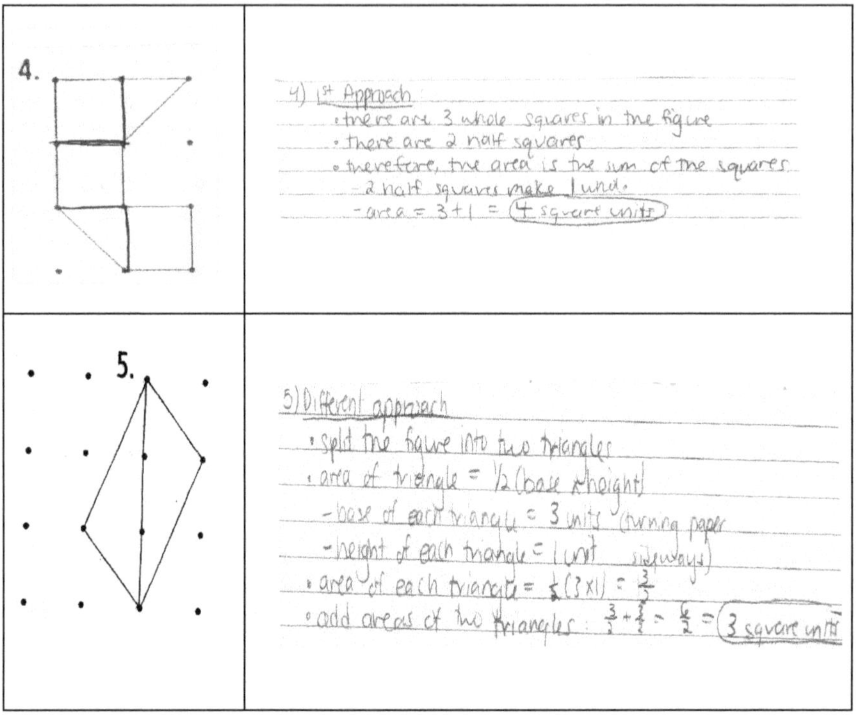

Figure 4. A Student's Description of Strategies Used to Find the Area

3.4 *Lesson debriefing with the teacher educator*

Immediately after the lessons, Mr. Orland, Miss Lipan, and teacher educator convened for an initial debriefing session to discuss observations about the level at which the students made use of their notes. A much longer debriefing session occurred the next day, which focused on how well Mr. Orland and Miss Lipan scaffolded students through the process, how effectively the reflection questions assisted students' examination of their notes to identify key ideas that helped them understand the content at hand, and how successful the model is in helping students manage information for later use. At this time, observations of the lesson were shared, which included specific thoughts about the mathematical ideas that the teachers modeled during the lessons. For example, Mr Orland's discussion focused on whether the launching and modeling of the problem helped students to understand how to translate the numerical data into a variable expression and whether students were able to examine the pattern as well as manipulate the data to yield an algebraic expression. We also discussed the extent to which students made use of their notes to appropriately produce the algebraic expression 2^{n-1} or 3^{n-1}.

The discussion with Miss Lipan centered on determining if the launching of the lesson should be the time at which students are provided with the definition of an irregular shape. She realized that she could have stated at the beginning of the modeling phase that, "Some shapes are not shaped like squares or triangles. They are irregular shapes. For example, many buildings are irregular." We also talked about whether or not too much emphasis was placed on the mechanics of carrying out the procedure. We concluded that because of the various learning disabilities represented in the classroom, it was necessary because the intent was for students to understand how to apply what they knew about finding the area of a square or a triangle. Furthermore, it was noted that as she recorded and modeled how to use the strategy, many students were taking notes. We concurred that the figure challenged some students because it was not immediately solvable. Repeating the steps during the review of the problem also helped students understand the procedures needed to arrive at an appropriate solution.

3.5 *Why does this model work?*

The section provides an explanation of how deliberate pedagogy serves as a scaffold for assisting students in learning how to take notes and then use these notes to solve mathematical problems. Most important, the reform-oriented curriculum places unusual demands on the students than that of a more traditional curriculum. It is imperative that these demands be taught to students who are compelled to meet them. Manouchehri and Goodman (2000) suggest that when using problem solving based curriculum, teachers need to include approaches that facilitate "guiding students' inquiry, mapping gradual development of both the content and learner's thinking, and creating a balance between fostering students' conceptual understanding while assisting them in acquisition of basic skills" (p. 29). According to the model described in this chapter, the teachers went beyond simply requiring students to take notes. Mr. Orland and Miss Lipan modeled their own thought processes through both writing and oral statements. These two modalities generate a quasi-interaction with students, in which students focus on capturing their teacher's application of an effective strategy and then restating in their own words what they observed and understood about the problem launched and modeled. This model allows both the teacher and students to make metacognitive advances; the teacher and students think about the way they are learning the mathematics at hand, as opposed to merely thinking about the mathematics content itself. This occurs because the teacher provides deliberate and explicit modeling of note taking, which helps the students to develop the metacognitive skills necessary for learning how to take notes that will subsequently help them to think about the concepts they are learning.

Metacognition refers to the abstract thought process through which an individual thinks about and reflects upon one's own thinking. In Miss Lipan's modeling, for example, she states, "I am going to name my strategy and prove how I can use it to find the area. My strategy is called *surround-and-subtract*." Then she both says and writes, "Step 1: Surround the original shape. Draw a square/rectangle around the borders of the original shape. Step 2: Find the total area of the square/rectangle you drew around the original shape. Step 3: Subtract the area of the outside shape from the total area of the square/rectangle you drew. This is Total Area – Outside shape = Area of Original Shape." In addition, Miss Lipan demonstrates how to solve the problem, making her implicit thinking and knowledge explicit for her students. Maccini, Mulcahy, and Wilson (2007) assert that, given the difficulties many students with learning disabilities in mathematics have accessing reform-based curriculum, it is essential to integrate pedagogical practices that are both deliberate and explicit.

Another reason that this model is an improvement over current use of note taking in mathematics instruction is that the teacher allows time for students to review and reflect on their notes, which are scaffolded by the written reflection questions. Giving crucial scaffolding is fundamental in the learning phase of note taking as learners progress through the zone of proximal development. Mr. Orland and Miss Lipan engage students in dialogue about the problem to provide any clarification that they might need about the task set before them. With respect to this approach, Farmer (1995) offers the explanation, "the ability to solve problems through dialogue with [teachers] or peers is a harbinger of competencies that will later become internalized" (p. 305). In this instance, the teachers also used many of the techniques that Rogoff (1990) describes as "scaffolded learning supports." These include the following: elaborating, linking, prompting, simplifying, and providing affective support.

An important observation is that because of organizing and applying a deliberate instructional approach, the actual situation may also be described as nominative. In other words, Mr. Orland and Miss Lipan set criteria pertaining to the most effective way of taking notes and, at the same time, state the requirements for meeting those criteria. The central point here is that the mathematics learning community, as in these two classrooms, exists as a means for appropriating deliberate discourse that is within reach for all learners. Students' note taking skills improve over time as students grow in their understanding of how their notes help them to organize, understand, and shape their ideas in a meaningful way. In addition to note taking as a deliberate part of instruction, teachers need to explicitly communicate their expectations regarding the use of written notes for completing a problem-solving task. These expectations must be generalized to classroom norms and procedures that are applied to all classroom activities; students must become accustomed to their teacher setting high expectations to prevent negative fallout that could result in low problem solving performance.

A final explanation of why this model works involves the collaborative nature of the model. As noted earlier, the teacher educator served as the more knowledgeable other in assisting Mr. Orland and Miss Lipan throughout the planning phase, observing and taking notes during the launching and modeling phase, and meeting with the teachers after the lesson to share observation notes, which included critical reflection and questions about the recent mathematics instruction. We worked together to achieve a shared understanding of the teaching and learning of mathematics. As Vygotsky (1978, 1981) theorizes, when teachers are challenged to work on activities collectively and are encouraged to achieve what they are not capable of doing individually, they are likely to move forward in their development as mathematics teachers, especially when the dialogue between the teacher and the other is sustained long enough to become a deliberate process. This, in turn, benefits students because it provides them with teachers that are more knowledgeable about how to make reform-oriented curricula accessible to all learners.

4 Conclusion

Bruner's conception of knowledge representation and Vygotsky's construct of the zone of proximal development have a place in contemporary discussion of the importance of creating effective learning communities that support all learners, especially when using reform-oriented curricula. Teachers need to understand how to assist all learners in their zone of proximal development, providing opportunities for problem-solving and inquiry-based activities that encourage the development of complex thinking and logical reasoning. Professional development activities that coalesce with Bruner and Vygotsky's research can assist teachers in the development of deliberate pedagogical practices. These practices must be designed to enhance mathematical learning experiences, such as supporting development of the students as effective note takers. Such a framework calls for a reconceptualization of the traditional role of teacher and learner. The emphasis is on processes and strategies rather than products and solutions. In other words, teachers must call attention to the *why* and *how* of mathematics, instead of merely focusing on the *what*—the final answer. This method of deliberate pedagogy allows teachers to move beyond the function of imparting knowledge and organizational skills. What discerns the note taking instructional approach described in this chapter is that it reflects the needs of the learner, the demands and purposes of the mathematical content, and the attributes of the context in which the launching and modeling transpired. As practicing teachers strive to work with diverse learners using reform-oriented mathematics curricula, the learner, the content, and the instructional context must be given considerable attention to be effective and successful (Albert, 2003; Draper, 2002; Maccini, Mulcahy, & Wilson 2007; White, 2003). Note taking can provide students with a structure for organization; further, when teachers deliberately use it to model the thinking and learning processes, note taking can provide much more—a metacognitive tool to enhance students' mathematical understanding.

References

Albert, L. R. (2003). Creating caring learning communities that support academic success. In C. Cimino, R. M. Haney, & J. M. O'Keefe, (Eds.). *Educating young adolescents: Conversations in excellence* (pp. 51-65). Washington, DC: National Catholic Education Association.

Albert, L. R. (2000). Outside in, inside out: Seventh grade students' mathematical thought process. *Educational Studies in Mathematics, 41*, 109-142.

Boch, F. & Piolat, A. (2005). Note taking and learning: A summary of research. *The WAC Journal, 16*, 101-113.

Boaler, J. (2002). Learning from teaching: Exploring the relationship between reform curriculum and equity. *Journal for Research in Mathematics Education, 33*(4), 239-258.

Bruner, J. S. (1996). *The culture of education.* Cambridge, MA: Harvard University Press.

Bruner, J. S. (1973). *The relevance of education.* New York: W. W. Norton.

Bruner, J. S. (1966). *Toward a theory of instruction.* New York: W. W. Norton.

Bruner, J. S. (1963). *On knowledge: Essays for the left hand.* Cambridge, MA: Harvard University Press.

Draper, R. J. (2002). School mathematics reform, constructivism, and literacy: A case for literacy instruction in the reform-oriented math classroom. *Journal of Adolescent and Adult Literacy, 45*(6), 520-529.

Driscoll, M. (1994). *Psychology of learning for instruction.* Boston: Allyn and Bacon.

Farmer, F. (1995). Voice reprised: Three etudes for a dialogic understanding. *Rhetoric Review, 13*, 304-320.

Greenfield, P. M. (1984). Theory of the teacher in the learning activities of everyday life. In B. Rogoff & J. Lave (Eds.), *Everyday cognition* (pp. 117-138). Cambridge, MA: Harvard University Press.

Lappan, G., Fey, J. T., Fitzgerald, W. M., Friel, S. N., & Phillips, E. D. (2004). *Connected mathematics; Growing, growing, growing*. Needham, MA: Pearson Prentice Hall.

Maccini, P., Mulcahy, C. A., & Wilson, M. S. (2007). A follow-up of mathematics interventions for secondary students with learning disabilities. *Learning Disabilities Research and Practice, 22*(1), 58-74.

Manouchehri, A. & Goodman, T. (2000). Implementing mathematics reform: Challenge within. *Educational Studies in Mathematics, 42*, 1-34.

National Council of Teachers of Mathematics. (2000). *Principles and standards for school mathematics*. Reston, VA: Author.

Paul, W. (1974). *How to study in college*. Boston: Houghton Mifflin.

Piolat, A. Olive, T., & Kellogg, R. (2005). Cognitive effort during note taking. *Applied Cognitive Psychology, 19*, 291-312.

Rogoff, B. (1990). *Apprenticeship in thinking: Cognitive development in social context.* New York: Oxford University Press.

Saye, J. W. & Brush, T. (2002). Scaffolding critical reasoning about history and social issues in multimedia-supported learning environments. *Educational Technology Research and Development, 50*(3), 77-96.

Simons, K. D. & Klein, J. D. (2007). The impact of scaffolding and student achievement levels in a problem-based learning environment. *Instructional Science, 35*, 41-72.

Vygotsky, L. S. (1994). The problem of cultural development of the child. In R. Van Der Veer and J. Valsiner (Eds.), *The Vygotsky reader* (pp. 57-72). Cambridge, Massachusetts: Blackwell.

Vygotsky, L. S. (1986). *Thought and language.* Cambridge, MA: MIT Press.

Vygotsky, L. S. (1981). The genesis of higher mental functions. In J. V. Wertsch (Ed.), *The concept of activity in Soviet psychology* (pp. 144-188). Armonk, NY: Sharpe.

Vygotsky, L. S. (1978). *Mind in society: The development of higher psychological processes*. Cambridge, MA: Harvard University.

Wertsch, J. V. (1980). The significance of dialogue in Vygotsky's account of social, egocentric, and inner speech. *Contemporary Educational Psychology, 5*, 120-162.

Wertsch, J. V. (1979). From social interaction to higher psychological processes: A classification and application of Vygotsky's theory. *Human Development, 22*, 1-22.

White, D. Y. (2003). Promoting productive mathematical classroom discourse with diverse students. *Journal of Mathematical Behavior, 22*, 37-53.

Chapter 5

Japanese Approach to Teaching Mathematics via Problem Solving

Yoshinori SHIMIZU

Japanese approach to teaching mathematics via problem solving is overviewed with a description of typical organization of mathematics lessons in Japanese schools. The selected findings of large-scale international studies of classroom practices in mathematics are examined for discussing the uniqueness of how Japanese teachers structure and deliver their lessons. The fundamental assumption that underlies the Japanese approach is discussed. In particular, how teachers plan a lesson by trying to allow mathematics to be problematic for students, to focus on the methods used to solve problem, and to tell the right things at the right times. Examples of textbook problems and anticipated students' solutions to them are presented to show how teachers share and analyze the solutions in the classroom discussion for achieving their goal of teaching mathematics. Finally, some practical ideas in the classroom shared by Japanese teachers are presented.

1 Introduction

Japanese mathematics teachers often organize an entire lesson by posing just a few problems with a focus on students' various solutions to them. They seem to share a belief that learning opportunities for their students are best raised when they are posed a challenging problem. Why do teachers in Japan consider teaching mathematics via problem solving beneficial? How do they achieve their goal of teaching mathematics content through the process of problem solving?

In this chapter, Japanese approach to teaching mathematics via problem solving is overviewed with a description of typical organization of mathematics lessons in Japanese classrooms. The fundamental assumption that underlies the Japanese approach is discussed. In particular, how teachers plan a lesson by trying to allow mathematics to be problematic for students, to focus on the methods used to solve problem, and to tell the right things at the right times. Examples of textbook problems and anticipated students' solutions to them are presented to show how teachers share and analyze the solutions in the classroom discussion for achieving their goal of teaching mathematics. Finally, some practical ideas in the classroom shared by Japanese teachers are presented.

2 Mathematics Lesson as Structured Problem-Solving

2.1 *A typical organization of a lesson*

Japanese teachers, in elementary (grades 1 to 6) and junior high (grades 7 to 9) schools, in particular, often organize an entire mathematics lesson around the multiple solutions to a few problems in a whole-class instructional mode. This organization is particularly useful when a new concept or a new procedure is going to be introduced during the initial phase of a teaching unit. Even during the middle or final phases of the teaching unit, teachers often organize lessons by posing a few problems with a focus on the various solutions students come up with.

A typical mathematics lesson in Japan, which lasts forty-five minutes in the elementary schools and fifty minutes in the junior high schools, has been observed to be divided into several segments (Becker, Silver, Kantowski, Travers, & Wilson, 1990; Stigler & Hiebert, 1999). These segments serve as the "steps" or "stages" in both the teachers' planning and delivery of the teaching-learning processes in the classroom (Shimizu, 1999):

- Posing a problem
- Students' problem solving on their own
- Whole-class discussion
- Summing up
- Exercises or extension (optional depending on time and how well students are able to solve the original problem.)

Lessons usually begin with a word problem in the textbook or a practical problem that is posed on the chalkboard by the teacher. After the problem is presented and read by the students, the teacher determines whether the students understand the problem well. If it appears that some students do not understand some aspect of the problem, the teacher may ask these students to read it again, or the teacher may ask questions to help clarify the problem. Also, in some cases, he or she may ask a few students to show their initial ideas of how to approach the problem or to make a guess at the answer. The intent of this initial stage is to help the students develop a clear understanding of what the problem is about and what certain unclear words or terms mean.

A certain amount of time (usually about 10-15 minutes) is assigned for the students to solve the problem on their own. Teachers often encourage their students to work together with classmates in pairs or in small groups. While students are working on the problem, the teacher moves around the classroom to observe the students as they work. The teacher gives suggestions or helps individually those students who are having difficulty in approaching the problem. He or she also looks for students who have good ideas, with the intention of calling on them in a certain order during the subsequent whole-class discussion. If time allows, the students who have already gotten a solution are encouraged by the teacher to find an alternative method for solving the problem.

When whole-class discussion begins, students spend the majority of this time listening to the solutions that have been proposed by their classmates as well as presenting their own ideas. Finally, the teacher reviews and sums up the lesson and, if necessary and time allows, then he or she poses an exercise or an extension task that will apply what the students have just learned in the current lesson.

2.2 *The Japanese lesson pattern*

The video component of the Third International Mathematics and Science Study (TIMSS) was the first attempt ever made to collect and analyse videotapes from the classrooms of national probability samples of teachers at work (Stigler & Hiebert, 1999). Focusing on the actions of teachers, it has provided a rich source of information regarding what goes on inside eighth-grade mathematics classes in Germany, Japan and the United States with certain contrasts among the three countries. One of the sharp contrasts between the lessons in Japan and those in the other two countries relates to how lessons were structured and delivered by the teacher. The structure of Japanese lessons was characterized as "structured problem solving", here again, while a focus was on procedures in the characterizations of lessons in the other two countries.

Table 1 shows the sequence of five activities described as the "Japanese pattern". In this lesson pattern, the discussion stage, in particular, depends on the solution methods that the students actually come up with. In order to make this lesson pattern work effectively and naturally, teachers need to have not only a deep understanding of the mathematics content, but also a keen awareness of the possible solution methods their students will use. Having a very clear sense of the ways students are likely to think about and solve a problem prior to the start of a lesson makes it easier for teachers to know what to look for when they are observing students work on the problem. The pattern seems to be consistent with the description of mathematics lessons as problem solving in the previous section, though there are some differences between them such as "reviewing the previous lessons" above and "exercises or extension" in the previous section.

Table 1
The Japanese lesson pattern (Stigler & Hiebert, 1999, pp.79-80)

Reviewing the previous lesson
Presenting the problems for the day
Students working individually or in groups
Discussing solution methods
Highlighting and summarizing the main point

2.3 *Beyond the pattern*

Characterization of the practices of a nation's or a culture's mathematics classrooms with a single lesson pattern was, however, problematised by the results of the Learner's Perspective Study (LPS) (Clarke, Mesiti, O'Keefe, Jablonka, Mok & Shimizu, 2007). The analysis suggested that, in particular, the process of mathematics teaching and learning in Japanese classrooms could not be adequately represented by a single lesson pattern for the following two reasons. First, lesson pattern differs considerably within one teaching unit, which can be a topic or a series of topics, depending on the teacher's intentions through out the sequence of lessons. Second, elements in the pattern themselves can have different meanings and functions in the sequence of multiple lessons. Needless to say, it is an important aspect of teacher's work not only to implement a single lesson but also to weave multiple lessons that can stretch out over several days, or even a few weeks, into a coherent body of the unit. It would not be possible for us to capture the dynamic nature of activities in teaching and learning process if each lesson was analysed as isolated.

An alternative approach was proposed to the international comparisons of lessons by the researchers in LPS team. That is, a postulated "lesson event" would be regarded to serve as the basis for comparisons of classroom practice internationally. In LPS, an analytical approach was taken to explore the form and functions of the particular lesson events such as "between desk instruction", "students at the front", and "highlighting and summarizing the main point" (Clarke, Emanuelsson, Jablonka & Mok, 2006).

In particular, the form and functions of the particular lesson event "highlighting and summarizing the main point", or "Matome" in Japanese, were analyzed in eighth-grade "well-taught" mathematics classrooms in Australia, Germany, Hong Kong, Japan, Mainland China (Shanghai), and the USA (Shimizu, 2006). For the Japanese teachers, the event "Matome" appeared to have the following principal functions: (i) highlighting and summarizing the main point, (ii) promoting students' reflection on what they have done, (iii) setting the context for introducing a new mathematical concept or term based on the previous experiences, and (iv) making connections between the current topic and previous one. For the teachers to be successful in maintaining these functions, the goals of lesson should be very clear to themselves, activities in the lesson as a whole need to be coherent, and students need to be involved deeply in the process of teaching and learning.

The results suggest that clear goals of the lesson, a coherence of activities in the entire lesson, active students' involvement into the lesson, are all to be noted for the quality instruction in Japanese classrooms. Also teachers need to be flexible in using a "lesson pattern", when they plan and implement a lesson as "structured problem-solving".

2.4 *A story or a drama as a metaphor for an excellent lesson*

Associated with the descriptions of "structured problem-solving" approach to mathematics instruction discussed above, several key pedagogical terms are shared by Japanese teachers. These terms reflect what Japanese teachers value in planning and implementing lessons within Japanese culture.

"Hatsumon", for example, means asking a key question to provoke and facilitate students' thinking at a particular point of the lesson. The teacher may ask a question for probing students' understanding of the topic at the beginning of the lesson or for facilitating students' thinking on a specific aspect of the problem. "Yamaba", on the other hand, means a highlight or climax of a lesson. Japanese teachers think that every lesson should include at least one "Yamaba". This climax usually appears as a highlight during the whole-class discussion. The point here is that all the activities, or some variations of them, constitute a coherent system called a lesson that hopefully includes a climax. Further, among Japanese teachers, a lesson is often regarded as a drama, which has a beginning, leads to a climax, and then invites a conclusion. Japanese teachers often refer to the idea of "KI-SHO-TEN-KETSU", which was originated in the Chinese poem, in their planning and implementation of a lesson. The idea suggests that Japanese lessons have a particular structure of a flow moving from the beginning ("KI", a starting point) toward the end ("KETSU", summary of the whole story).

If we take a story or a drama as a metaphor for considering an excellent lesson, a lesson needs to have a highlight or climax based on the active role of students guided by the teacher in a coherent way. Stigler and Perry (1988) found *reflectivity* in Japanese mathematics classroom. They pointed out that the Japanese teachers stress the process by which a problem is worked and exhort students to carry out procedure patiently, with care and precision. Given the fact that the schools are part of the larger society, it is worthwhile to look at how they fit into the society as a whole. The reflectivity seems to rest on a tacit set of core beliefs about what should be valued and esteemed in the classroom. As Lewis noted, within Japanese schools, as within the larger Japanese culture, Hansei—self-critical reflection—is emphasized and esteemed (Lewis, 1995).

In sum, the selected findings of large-scale international studies of classroom practices in mathematics examined above suggest that "structured problem solving" in the classroom with an emphasis on students' alternative solutions to a problem can be a characterization of Japanese classroom instruction from a teacher's perspective. Also, a coherence of the entire lesson composed of several segments, students' involvement in each part of the lesson, and the reflection of what they did are all to be noted for the approach taken by Japanese teachers. To comprehend what Japanese teachers value in their instruction with a cultural bias, a story or a drama can be a metaphor for characterizing an excellent lesson in Japan.

3 Preparing a Lesson by Focusing on Students' Problem Solving

To prepare a lesson with a focus on students' problem solving, teachers need to plan it by trying to allow mathematics to be problematic for students, to focus on the methods used to solve problem, and to tell the right things at the right times. The following example illustrates these points.

Here is a typical construction problem of an angle bisector (*see Figure 1*). The topic is taught in 7th grade within the current national curriculum standard in Japan. Students are expected to learn how to draw the bisector to any given angle by using compass and straightedge.

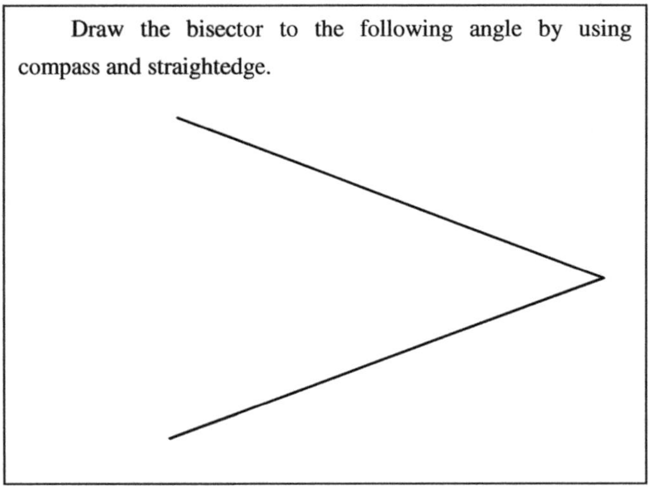

Figure 1. The angle bisector problem

To make the mathematics problematic for students, the teacher first needs to pose a thought-provoking, but not too difficult, problem for the students to solve. As for the angle bisector problem, teacher can use a paper on with the problem is printed and tear it off in front of the students as shown in Figure 2.

When you were going to draw the bisector to the angle by using compass and straightedge, the paper was torn off as below. Can you still draw the bisector to the original angle?

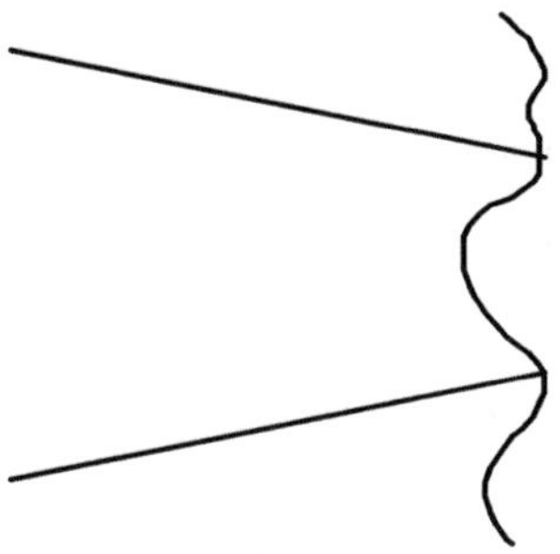

Figure 2. The angle bisector problem revised

By setting the context for the angle as described above, students are involved in the problem situation and will start to think about it deeply. Students will come up with various methods by using mathematical ideas they have learned. The "torn- off" angle bisector problem can be used in the classroom of 8th or 9th grade students. Figures 3a, 3b, and 3c show some of the solutions that were found by the students in 9th grade mathematics classroom. Figures 3a and 3b are the solutions by using inscribed circle to the given lines in different ways. The student who produced Figure 3c used parallel line (l3) to have triangle ABC. By extending the line BC to get point D on the other given line (l2), we can see the large "triangle". The angle bisector line is drawn as the perpendicular line to the "bottom" BD.

In the lesson, certain amount of time is to be assigned for the students to solve this challenging problem on their own. Teachers may encourage their students to work together with classmates in pairs or in small groups, if they have difficulty understanding it. While students are working on the problem, the teacher needs to observe the students as they work. The teacher may give suggestions or help individually those students who are having difficulty in approaching the problem and also look for the students who have good ideas with the intention of calling

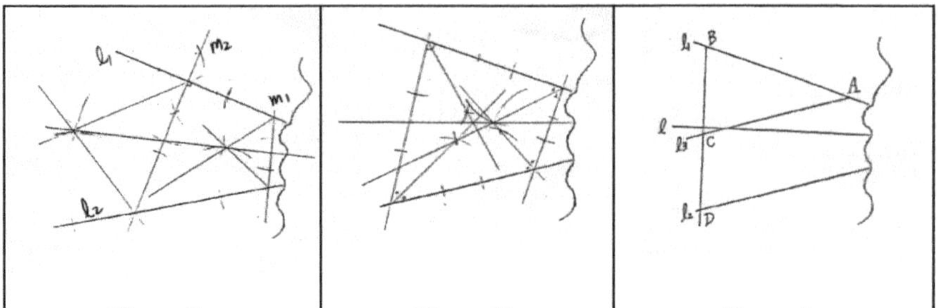

Figure 3a | Figure 3b | Figure 3c

Figure 3. Students' alternative solutions to the problem

on them in a certain order during the subsequent whole-class discussion. The students who have already gotten a solution should be encouraged by the teacher to find an alternative method for solving to the problem.

In a whole-class discussion, students' solutions are presented and discussed. The focus here is not just presenting alternative solutions but to reflect on them to consider similarities and differences among the methods from mathematical points of view. There are many ideas used to solve the problem. The mathematical ideas used to solve the problem can be classified into groups and then integrated. Through the discussion the students can understand that the key to the solutions can be regarded as transforming the "missing angle" to appear again on the paper. Before implementing the lesson, teachers need to think about aspects described above. Throughout the process of preparation and implementation of a lesson, teachers need to analyze the topic carefully in accordance with the objective(s) of a lesson. The analysis includes analyses of the mathematical connections both between the current topic and previous topics (and forthcoming ones in most cases) and within the topic, anticipation for students' approaches to the problem presented, and planning of instructional activities based on them.

4 Some Practical Ideas Shared by Japanese Teachers

Various teachers with whom I have worked over the past several years have made numerous suggestions to me regarding the Japanese approach to teaching mathematics (Shimizu, 2003). Among these suggestions, five are especially pertinent to the focus of this chapter.

Suggestion 1: Label students' methods with their names

During the whole-class discussion of the students' solution methods, each method is labeled with the name of the student who originally presented it. Thereafter, each solution method is referred by the name of student in the discussion. This practical technique may seem to be trivial but it is very important to ensure the student's "ownership" of the presented method and makes the whole-class discussion more exciting and interesting for the students.

Suggestion 2: Use the chalkboard effectively

Another important technique used by the teacher relates to the use of the chalkboard, which is referred as "bansho" (board writing) by Japanese teachers. Whenever possible, teachers put everything written during the lesson on the chalkboard without erasing. By not erasing anything the students have done and placing their work on the chalkboard in a logical, organized manner, it is much easier to compare multiple solution methods. Also, the chalkboard can be a written record of the entire lesson, giving both the students and the teacher a birds-eye view of what has happened during the lesson.

Suggestion 3: Use the whole-class discussion to polish students' ideas

The Japanese word, "neriage," is used to describe the dynamic and collaborative nature of a whole-class discussion in the lesson. This word, which can be translated as "polishing up", works as a metaphor for the process of "polishing" students' ideas and getting an integrated mathematical idea through a dynamic whole-class discussion. Japanese teachers regard "neriage" as critical to the success or failure of the entire lesson.

Suggestion 4: Choose the context of the problem carefully

The specific nature of the problem presented to the students is very important. In particular, the context for the problem is crucial for the students to be involved in it. Even the numbers in word problems are to be carefully selected for eliciting a wide variety of student responses. Careful selection of the problem is the starting point for getting a variety of student responses.

Suggestion 5: Consider how to encourage a variety of solution methods

What else should the teacher do to encourage a wide variety of student responses? There are various things the teacher can do when the students come up with only a few solution methods. It is important for the teacher to provide additional encouragement to the students to find alternative solution methods in addition to their initial approaches.

5 Final Remarks

The Japanese approach to teaching mathematics via problem solving usually takes a form of organizing an entire lesson around posing one or two problems with a focus on the subsequent discussion of various solution methods generated by the students. The students' own ideas are incorporated into the classroom process of discussing multiple solution methods to the problem. In this approach problem solving is an essential vehicle for teaching mathematics. This instructional approach is not used only on special occasions or once per week. Rather, it is the standard approach followed for teaching ALL mathematics content.

In order for lessons to be successful, teachers have to understand well the relationship between mathematics content to be taught and students' thinking about the problem to be posed. Anticipating students' responses to the problem is the crucial aspect of lesson planning in the Japanese approach to teaching mathematics through problem solving.

References

Becker, J.P., Silver, E.A., Kantowski, M.G., Travers, K.J., & Wilson, J.W. (1990, October). Some observations of mathematics teaching in Japanese elementary and junior high schools. *Arithmetic Teacher, 38*, 12-21.

Clarke, D., Emanuelsson, J., Jablonka, E., & Mok, I.A.C., (Eds.). (2006). *Making connections: Comparing mathematics classrooms around the world.* Rotterdam: Sense Publishers.

Clarke, D., Mesiti, C., O'Keefe, C., Jablonka, E., Mok, I.A.C., & Shimizu, Y. (2007). Addressing the challenge of legitimate international comparisons of classroom practice. *International Journal of Educational Research, 46*, 280-293.

Lewis, C. (1995). *Educating hearts and minds: Reflections on Japanese preschool and elementary education.* New York: Cambridge University Press.

Shimizu, Y. (1999). Aspects of mathematics teacher education in Japan: Focusing on teachers' role. *Journal of Mathematics Teacher Education, 2*(1), 107-116.

Shimizu, Y. (2003). Problem solving as a vehicle for teaching mathematics: A Japanese perspective. In F.K. Lester (Ed.,), *Teaching mathematics through problem solving: Grades Pre K - 6*, (pp. 205-214). Reston, VA: National Council of Teachers of Mathematics.

Shimizu, Y. (2006). How do you conclude today's lesson? The form and functions of "Matome" in mathematics lessons. In D. Clarke, J. Emanuelsson, E. Jablonka & I. A. C. M. (Eds.) *Making Connections: Comparing Mathematics Classrooms Around the World* (pp. 127-145). Rotterdam: Sense Publishers.

Stigler, J.W., & Hiebert, J. (1999). *The teaching gap: Best ideas from the world's teachers for improving education in the classroom.* New York: NY, The Free Press.

Stigler, J.W., & Perry, M. (1988). Cross cultural studies of mathematics teaching and learning: Recent findings and new directions. In D.A. Grouws, & T.J. Cooney (Eds.) *Perspectives on research on effective mathematics teaching* (pp. 194-223). Mahwah, NJ.: Lawrence Erlbaum Associates & Reston, VA: National Council of Teachers of Mathematics.

Chapter 6

Mathematical Problem Posing in Singapore Primary Schools

YEAP Ban Har

Mathematical problem posing is the generation of mathematics problems as well as the reformulation of existing ones. The chapter focuses on problem-posing process that primary school students engaged in. These processes include (a) posing *primitives*, (b) posing related problems, (c) constructing meaning for a mathematical operation, (d) engaging in metacognition, and (e) connecting to one's experiences. In this chapter, examples on the use of problem posing before, during and after problem solving are given and illustrated. The use of problem posing for various instructional goals such as to develop concepts, for drill-and-practice, for problem solving, to assess understanding, and to provide differentiated instruction is described and illustrated using cases from Singapore schools.

1 Introduction

Mathematical problem posing is defined as the generation of new problems, as well as the reformulation of existing ones (Silver, 1994). Silver delineated three types of problem posing, namely problem posing that occurs before, during or after problem solving. Silver's definition suggests that it is necessary to consider mathematical problem posing in any discussion on mathematical problem solving.

In Singapore, the mathematics curriculum framework focuses on mathematical problem solving. Among the aims of the curriculum, it is hoped that students are able to "formulate and solve problems (p. 5, Ministry of Education, 2006a; p.1, Ministry of Education 2006b.)". The other chapters in this book focus on various aspects of mathematical problem solving. This chapter focuses on mathematical problem posing. Problem-posing tasks are also common in primary school textbooks used in Singapore (e.g. Fong, Ramakrishnan & Gan, 2007).

The benefits of using problem-posing tasks in the mathematics classrooms has been investigated across grade levels and cannot be ignored as such tasks can influence, among other things, students' (1) aptitude in mathematics, including understanding and problem-solving ability, (2) attitudes towards mathematics, including curiosity and interest, and (3) ownership of their work (English, 1997a; Grundmeier, 2002; Knuth, 2002; Perrin, 2007).

In the first part of the chapter, some research on mathematical problem posing are presented. In particular, selected research on the relationship between problem solving and problem posing are described. The findings from one research on Singapore students focusing on the problem-posing processes are presented. In the second part of the chapter, the different roles of problem-posing tasks in the classroom are described.

2 Mathematical Problem Solving and Problem Posing

The relationship between mathematical problem solving and problem posing has been the subject of many research studies. Students who were better in non-routine problem solving were better problem posers. Silver and Cai (1996) found that problem-solving ability of American middle school students highly correlated with their ability to pose semantically complex problems in one type of problem-posing task. In a series of investigations on third, fifth and seventh graders in Australia, English (1997b, 1997c, 1998) found some relationship between problem solving and problem posing. In particular, she found that competence in routine problem solving is associated with posing of computationally complex, but not necessarily structurally complex, problems. Competence in novel problem solving is associated with posing of structurally complex problems. Among Singapore students, it was found that good problem solvers had significantly higher problem-posing scores than poor problem solvers (Yeap, 2002). In addition, it was found that when the students had no prior experience in problem posing, the relationship between problem solving and problem posing was not dependent on grade level. In all these research studies, students were asked to pose the problems given some stimulus. Research on problem posing during and after problem solving is comparatively less established.

3 Mathematical Problem-Posing Processes

Kilpatrick (1987) argued that one of the basic cognitive processes involved in mathematical problem posing is association, which was confirmed in a study by Silver and Cai (1996). Winograd (1990) found that many students generated and selected information from their experiences or immediate physical environment when they posed problems. More recently, Christou, Mousoulides, Pittalis, Pantazi and Sriraman (2005) studied sixth graders to understand processes that students used during mathematical problem posing. Among others, it was found that the students were engaged in selecting quantitative information during problem posing.

In a study on third grade and fifth grade Singapore students posing arithmetic word problems, it was found that primary school students engaged in five categories of problem-posing processes (Yeap, 2002). The categories included:

1. posing *primitives*,
2. posing related problems,
3. constructing meaning for a mathematical operation,
4. engaging in metacognition, and
5. connecting to one's experiences.

3.1 *Posing primitives*

According to Krutetskii (1976), students who are good in mathematics can see the 'hidden' questions when presented with text containing numerical information. For example, in a test to identify mathematically-able students, the test requires students to pose a question that follows the text "25 pipes of lengths 5 m and 8 m were laid over a distance of 155 m." The unstated question is "How many pipes of each kind were laid?" Silver and Cai (1996) referred to such questions as *primitives*.

Figure 1 shows a problem-posing task used in the study to investigate problem-posing processes (Yeap, 2002). The first statement is a relational one which compares the number of girls in two classes. The second statement is an assignment one which describes the number of boys in the two classes.

The students' responses were of four types (see Table 1). Many students posed questions to determine the number of boys in the two classes. Responses in Category A were common as were responses in Category B, where students posed questions to determine the number of

> There are 3 more girls in Primary 4A than in Primary 4B.
> There are 15 boys in each class.
>
> Write three mathematics questions about the two classes.
> You can include other numbers, if you like.

Figure 1. A Task for Mathematical Problem Posing (Task 1)

girls in Primary 4A, given the number of girls in Primary 4B. Significantly fewer students posed questions to determine the number of girls in Primary 4B, given the number of girls in Primary 4A. Similarly, significantly fewer students posed questions to determine the number of girls in one class, given the number of children in the class, or to determine the number of children in one class, given the number of girls in that class.

Table 1
Responses to Task 1

Category of Response	Sample Responses
A	*How many boys are there in both classes?*
B	*How many girls are there in Primary 4A if there are 33 girls in Primary 4B?*
C	*If there are 40 girls in Primary 4B, how many girls are there in Primary 4A?*
D	*Miss Sem teach 4A. 4A has 27 girls. How many pupils are there in 4A?* *If there are 40 pupils in Pr. 4A, how many girls are there (in 4A)?*

Note: The sample responses were taken verbatim from the data (Yeap, 2002).

3.2 *Posing related problems*

Kilpatrick (1989) argued that one of the basic cognitive processes involved in problem posing is making associations. Silver and his associates have previously explored this process in several studies (Leung, 1993; Silver & Cai, 1996; Silver et al., 1996).

Using the task shown in Figure 1, Yeap (2002) found that students posed three types of related questions to a given situation. Two of the three types of relatedness are described here. The first type of related questions is called *serial questions*. Students are said to have posed serial questions when each question requires information from the previous one. Figure 2 shows two such responses.

Response 1

How many girls are there in Primary 4A?

How many girls are there in Primary 4B?

Response 2

If there are 24 girls in 4B, how many girls are there in 4A?

How many students are there in 4A?

How many students are there in the two classes?

Figure 2. Serial questions

Response 1

If there are 20 girls in Primary 4A. How many pupils are there in Primary 4B.

If there are 33 pupils in Primary 4A. How many pupils are there in Primary 4B.

Response 2

There are 30 children in 4A. How many children are there in 4B?

There are 32 children in 4B. How many children are there in 4A?

There are 21 girls in 4A. How many girls are there in 4B?

Figure 3. Parallel questions

The second type of related questions is called *parallel questions*. Students are said to have posed parallel questions when each of the questions have the same structure. The answer of the preceding question does not facilitate the answering of the one that follows. Figure 3 shows two such responses.

3.3 *Constructing meaning for mathematical symbols*

Yeap (2002) found that when students were asked to pose a word problem that has as one of its solution steps the multiplication sentence 4×6, they posed problems to show the meanings of multiplication that they had been taught. For example, none of the third graders posed problems involving area as they had not been taught the concept. Neither the third graders nor the fifth graders had any experiences with combinatorial problems, hence none of them posed such problems. Although it was expected that equal groups problem would have been easier, it was found that the students were just as likely to pose the more difficult multiplicative comparison type problems. This could be because the Singapore third grade textbooks emphasize the more difficult multiplicative comparison and rate problems. Previous studies with students in the same age group (English, 1996) found that students tended to pose equal group problems and few of them posed multiplicative comparison problems. Table 2 shows examples of the students' responses.

Table 2
Meanings Students Associated Multiplication with

Type of Situation	Sample Responses
Equal Groups	*There are 4 boxes of oranges. Each box, there are 6 oranges. How many oranges are there?*
Comparison	*Tom collected 6 phone cards. I collected 4 times as many phone cards as him. How many phone cards did I collected?*
Rectangular Array	*There are 6 chairs in a row. If there are 4 rows how many chairs were there?*
Area	*A rectangular room has a breadth o4 m and a length 6 m. What is the area of the rectangular room?*
Rate	*Ahmad bought a box for $4. He needs to buy another 5 more boxes. How much money would he have spent for all the boxes?*

3.4 *Engaging in metacognition*

A few examples of students engaging in metacognition as they posed problems are given below. A student demonstrated substantial monitoring of his problem-posing process from the way he edited his problem as he posed it. When asked to pose a problem that has an answer of 10, he initially wrote: *Mary has $60.* which he edited to *Mary has $30.* He then continued: *Ali has* $\frac{1}{6}$ *of Mary's money.* And included a third person, probably to make the problem more complex: *if Jane and Mary have the same amount of money.* However, he changed *Mary* to *Ali* and his eventual statement was: *if Jane and Ali have the same amount of money, what is ...* At this point, he edited his text because he probably realized the answer would not be 10. He deleted his pending question and continued: *and peter has spent all his money how much is their Average money altogether?*

Another student wrote: *Britney Spears has 100 balloons. 10 of them burst. 20 of them flied away. 30 of them had been stole and 30 of them had been given away.* to satisfy the condition that the answer must be 10. Her initial question was: *How many balloons had been stole than been given away?* She decided to make hers a two-part problem and wrote the second part: *How many balloons are there left?* The student demonstrated ability to monitor her thoughts as she went back to her original text and changed two numbers. Her final text read: *Britney Spears has 100 balloons. 10 of them burst. 20 of them flied away. <u>40</u> of them had been stole and <u>20</u> of them had been given away.*

3.5 *Connecting to one's experiences*

Ellerton (1986) found that the content and style of students' problems uniquely reflect their mathematical experiences and ideas. Menon (1995) found that children tended to pose problems based on their non-mathematical experiences. Yeap (2002) found that students in Singapore were making connections to their non-mathematical experiences as well as to their experiences with textbook problems.

A primary five student used the names of real people in his problems. All the names he included were his good friends' and his. He also used an object that he was probably familiar with (a popular toy called *pokemon*) in some of his problems. He was evidently using his non-mathematical experiences when he posed his problems. The effect was, however, on superficial features of the problems. Figure 4 shows an example of the problems posed by this student.

> *Qiwen has 25 pokemon. Iman has 10 pokemon and Huang Yong has 5 pokemon. Iman and Huang Yong decided to combine their pokemon together. What is the difference of Iman and Huang Yong pokemon and Qiwen's pokemon.*

Figure 4. Iman's problem

In a rare case, a student used her knowledge in another school subject to pose her problems. She wrote a problem based on science facts she knew, that an insect has 6 legs and that an ant is an insect. She wrote: *There are 4 ants. Each ant has 6 legs. How many legs are there altogether?*

The textbooks seemed to have a big influence on the problems posed by the students. Among the primary three students, it was surprising that more of them posed comparison problems (*Tom has 4 stamps. Kelvin has 6 times as much as Tom. How many stamps did Kelvin have*?) than equal group problems (*There are 4 boxes oranges. Each box, there are 6 oranges. How many oranges are there?*) although the former is structurally more challenging. Textbook analysis revealed that the primary three textbook emphasized the comparison problems when dealing with multiplication. There were 16 comparison problems and eight equal group problems in the textbooks used by the students in the study.

The textbooks also influenced the students in more superficial ways. Many students used names commonly used in the textbook word problems such as John and Mary, although these names were rare among the peers of the students in the study. For example, John was the character in two word problems among nine that appeared on two pages of a textbook. And children would pose problems about John and Mary: *John has 11 sweets. If he eats 1 sweet. How many will he has now? Mary has 956 stickers. John has 326 stickers more than Mary. How many stickers do they have altogether? And John has 5 balls. Mary has twicw as many as John. How many ball had Mary?*

In modelling their word problems after textbook ones, some students suspended their sense of reality. A primary five students, when asked to pose a problem that has an answer of 4 × 6, wrote *John is 6 m tall and his sister is 4 times as tall as him. How tall is the sister?*. Others ignored relationships. A primary three student posed this problem: *There are 12 adults in the hall. There are 3 times as many women as adults. How many women are there?*. The student failed to see that women are a subset of adults.

The preceding sections describe some processes that students engage in when they pose mathematical problems. In the next section, the roles of mathematical problem posing in the classroom are discussed.

4 Mathematical Problem Posing in the Classroom

The use of mathematical problem posing to develop concepts, to provide drill-and-practice, for problem solving, as an assessment tool, as a motivational tool and to cater to mixed-ability classes are described with specific examples.

4.1 *Developing a concept*

Mrs Pang showed a class two rectangles and asked the primary three students to ask questions about the two rectangles. Among the questions posed was one about the relative size of the two rectangles: Which rectangle is bigger?

Based on the question posed by students, Mrs Pang conducted a lesson on the use of square tiles to help students develop the concept of area of a figure.

The use of problem posing helps students achieve a focus in a lesson. The question posed, Which rectangle is bigger?, became the focus of the lesson on area. When problem posing is used on a regular basis, students also develop the ability to focus on significant aspects of situations presented to them.

4.2 *Providing drill-and-practice*

Mr Osman asked his primary six students to sketch composite figures that included circles or part of circles such that the area of each figure was 154 cm^2 instead of asking students to compute the areas of composite figures drawn for them. The students were allowed the use of calculators.

In coming up with the required figures, students practiced repeatedly the use of formulae to calculate area of various figures including circles. In addition, students had the opportunity to evaluate if their figures satisfied the given conditions. They also had the opportunity to make adjustments to 'incorrect' figures to obtain the required ones. The students also had the chance to exercise their creativity and tried to out-do each other by coming up with figures that none of their peers had come up with.

The use of problem posing allows teachers to add value to drill-and-practice activities by engaging students in a range of higher-order thinking skills and habits of mind.

4.3 *Problem solving*

Miss Siti asked her primary six students to pose questions based on the text of a word problem she was using to teach problem solving in the topic of speed. The students were given the text shown in Figure 5.

> David and Michael drove from Town A to Town B at different speeds. Both did not change their speeds throughout their journeys. David started his journey 30 minutes earlier than Michael. However, Michael reached Town B 50 minutes earlier than David. When Michael reached Town B, David had travelled $\frac{4}{5}$ of the journey and was 75 km away from Town B.

Figure 5. Text used in Miss Siti's class

Some of the questions asked by the students are:

- What was the distance between the two towns?
- Who took more time for the journey? How much more?
- How much time did David take? How much time did Michael take?
- What was David's speed? What was Michael's speed?

Subsequently, Miss Siti asked students to decide the questions that can be answered directly using the information in the text and those that require further information before they can be answered.

In solving a complex, multi-step problem, students have to know the intermediate questions they need answers to. By giving students, many opportunities to pose problems in problem-solving lessons, teachers are essentially teaching them the problem-solving process.

4.4 *Assessing understanding*

Mr Iqbal asked his primary three students to make up three word problems that can be solved by doing 4×6. He also encouraged them to make the problems as different as possible.

He was able to assess his students conceptual understanding of multiplication by looking at the situations the students used in the word problems. Students who lacked an understanding of multiplication posed problems such as *Ani has 4 sweets and Bala has 6 sweets. How many sweets do they have altogether?* and David had 20 books and he bought another 4 books. *How many books did David have after buying the 4 books?*

Students who had the appropriate concept of multiplication posed problems such as: *Bob had 6 bags of flowers each bag of flower had 4. How many does she had.* and *Tom collected 6 phone cards. I collected 4 times as many phone cards as him. How many phone cards did I collected?*

Mr Iqbal was also able to see that some of his students were more advanced as they posed problems with situations that most primary three students do not often associate with multiplication. Such students posed problems involving array (*There are 6 chairs in a row. If there are 4 rows how many chairs were there?*) and rate (*Mr Tan bought 6 kg of fish. Each fish costs $4 each. How much did he pay?*)

By using a problem-posing task Mr Iqbal was able to go beyond assessing procedural knowledge. He was able to get a glimpse of the students' conceptual understanding.

4.5 *Differentiating instruction*

Madam Gowri likes to use problem-posing in her mixed-ability class. In asking her primary four students to write a word problem that includes the numbers $\frac{2}{3}$, 6 and 24, she allowed her struggling students to leave out one of the whole numbers and encouraged her advanced students to make their problems multi-step. She asked the latter to also make their problems more challenging.

This allows the struggling students to handle the tasks at their level and that in itself is motivating. Problem posing also prevents the advanced students from becoming bored with standard tasks as they were able to challenge themselves by trying to figure out how to compose the three numbers in a way that the problem is solvable. While the average student may pose a problem such as: *Primary 3A has 6 boys and 24 girls. $\frac{2}{3}$ of the students say they like pizza. How many students like pizza?* An advanced student may pose problem: *John read 6 pages of a book on Monday. He read $\frac{2}{3}$ of the remaining pages on Tuesday and still has 24 pages left. How many pages does the book have?*

Miss Gan often uses what-if questions with her students. When she asked her primary three students to use the digits 0 to 9 exactly once to make a correct addition sentence, some managed to do it quickly while others struggled for a long time.

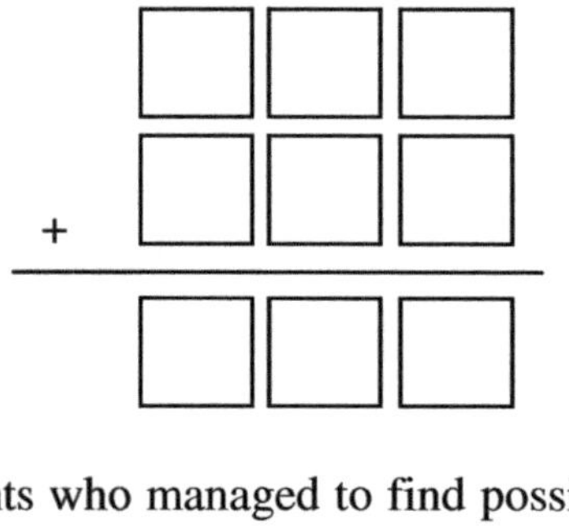

She asked students who managed to find possible solutions quickly to ask themselves what-if questions. Some of them asked what if the sum is a four-digit number. Other asked what if they were not allowed to use, say, the digit 0. Miss Gan then asked these students to solve the problem based on the what-if questions. This provided advanced students to be engaged with challenging, self-directed tasks which in turn was a form of motivation.

5 Conclusion

This chapter outlines some processes that students engage in when they pose mathematical problems. An understanding of these processes allows teachers to choose appropriate problem-posing tasks for classroom use. Different types of problem-solving tasks including problem posing before problem solving (such as Mr Iqbal's example), problem posing during problem solving (such as Miss Siti's example) and problem posing after problem solving (such as Miss Gan's example) illustrates the various roles that problem-posing tasks can play in the classroom.

Some ideas for classroom research include investigating the problems students posed as well as investigating the effects of problem posing as an intervention. Investigations into problems posed by students can be used to explore students' mathematical understand as well as generic ability such as creativity. Investigations using problem posing as an intervention can show it effects on problem solving and attitudes.

References

Ellerton, N. F. (1986). Children's made up problems: A new perspective of talented mathematicians. *Educational Studies in Mathematics, 17*, 261-271.

English, L. D. (1996). Children's problem-posing and problem-solving preferences. In J. Mulligan & M. Mitchelmore (Eds.), *Children's number learning* (pp. 227-242). Adelaide, Australia: The Australian Association of Mathematics Teachers Inc.

English, L. D. (1997a). Promoting a problem-posing classroom. *Teaching Children Mathematics, 4*, 172-179.

English, L. D. (1997b). The development of fifth-grade children's problem-posing abilities. *Educational Studies in Mathematics, 34*(3), 183-217.

English, L. D. (1997c). Seven-grade students problem posing from open-ended situations. In F. Biddulph & K. Carr (Eds.), *People in Mathematics Education* (pp. 39-49). Sydney: Mathematics Education Research Group of Australasia Inc.

English, L. D. (1998). Children's problem posing within formal and informal context. *Journal of Research for Mathematics Education, 29*(1), 82-106.

Fong, H. K., Ramakrishnan, C., & Gan, K. S. (2007). *My Pals Are Here! Maths Second Edition*. Singapore: Marshall Cavendish Education.

Grundmeier, T. A. (2002). University students' problem-posing abilities and attitudes towards mathematics. *PRIMUS, 12*, 122-134.

Kilpatrick, J. (1987). Problem formulating: Where do good problems come from? In A. H. Schoenfeld (Ed.), *Cognitive science and mathematics education* (pp. 123-147). Hillsdale, NJ: Lawrence Erlbaum.

Knuth, E. J. (2002). Fostering mathematical curiosity. *Mathematics Teacher, 95*, 126-130.

Krutetskii, V. A. (1976). *The psychology of mathematical abilities in school children.* Chicago: The University of Chicago Press.

Leung, S. S. (1993). *The relations of mathematical knowledge and creative thinking to the mathematical problem posing of prospective elementary school teachers on tasks differing in numerical content.* Unpublished doctoral dissertation, University of Pittsburg.

Menon, R. (1995). The role of context in student-constructed questions. *Focus on Learning Problems in Mathematics, 17*(1), 25-33.

Ministry of Education (2006a). *Mathematics syllabus: Primary*. Singapore: Curriculum Planning and Development Division.

Ministry of Education (2006b). *Mathematics syllabus: Secondary*. Singapore: Curriculum Planning and Development Division.

Perrin, J. R. (2007). Problem posing at all levels in the calculus classroom. *School Science and Mathematics, 107*(5), 182-192.

Silver, E. A. (1994). On mathematical problem posing. *For The Learning of Mathematics, 14*, 19-28.

Silver, E. A. & Cai, J. F. (1996). An analysis of arithmetic problem posing by middle school students. *Journal of Research for Mathematics Education, 27*(5), 521-539.

Silver, E. A., Mamona-Downs, J., Leung, S. S., & Kenney, P. A. (1996). Posing mathematical problems: An exploratory study. *Journal of Research for Mathematics Education, 27*(3), 293-309.

Winograd, K. (1990). *Writing, solving and sharing original math story problem: Case studies of fifth grade children's cognitive behavior.* Unpublished doctoral dissertation, University of Northern Colorado.

Yeap, B. H. (2002). *Relationship between children's mathematical word problem posing and grade level, problem-solving ability and task type.* Unpublished doctoral dissertation, Nanyang Technological University.

Chapter 7

Solving Mathematical Problems by Investigation

Joseph B. W. YEO YEAP Ban Har

Most educators would think of heuristics when it comes to solving closed mathematical problems, while many researchers believe that mathematical investigation must be open and is different from problem solving. In this chapter, we discuss the relationship between problem solving and investigation by differentiating investigation as a task, as a process and as an activity, and we show how the process of investigation can occur in problem solving if we view mathematical investigation as a process consisting of specialising, conjecturing, justifying and generalising. By looking at two examples of closed mathematical tasks, we examine how investigation can help teachers and students to solve these problems when they are stuck and how it can aid them to develop a more rigorous proof for their conjectures. We also deliberate whether induction is proof and how heuristics are related to investigation. Finally, we consider the implications of the idea of solving mathematical problems by investigation on teaching.

1 Introduction

The use of problem-solving heuristics or strategies to solve mathematical problems was popularised by Pólya (1957) in his book *How to solve it* (first edition in 1945). Few educators would talk about solving mathematical problems by investigation. In fact, many educators (e.g., HMI, 1985; Lee & Miller, 1997) believe that mathematical investigation must be open and that it must involve problem posing. Thus the idea of solving closed mathematical problems by investigation is a contradictory notion. Although many educators (e.g., Evans, 1987; Orton & Frobisher, 1996) have observed that there are overlaps between problem solving and investigation, they usually ended up separating them as distinct processes: problem solving is convergent while investigation is divergent (HMI, 1985). Some educators (e.g., Pirie, 1987) have even claimed that it is not fruitful to discuss the similarities and differences between them, but we agree with Frobisher (1994) that this is a crucial issue that may affect how and what teachers teach their students. Therefore, the main purposes of this chapter are to clarify the relationship between problem solving and investigation, to illustrate how investigation can help teachers and students to solve two closed mathematical problems when they are stuck, and to discuss how they can make use of investigation to develop a more rigorous proof for their conjectures.

We begin by examining what constitutes a problem to a particular person, whether problems must be closed or whether they can be open, and how investigation is related to problems. Subsequently, we discuss the relationship between investigation and problem solving by first separating investigation into investigative tasks, investigation as a process and investigation as an activity, and then characterising the process of mathematical investigation as involving the four core thinking processes of specialising, conjecturing, justifying and generalising. We argue that investigation as a process can occur when solving closed mathematical problems and we examine how investigation can aid teachers and students to solve these problems when they are stuck by looking at two closed mathematical tasks. In particular, we observe how investigation can help them to develop a more rigorous proof for their conjectures. Then we deliberate whether induction is proof by looking at the different meanings of the terms ‘induction’, ‘inductive observation’ and ‘inductive reasoning’, and we consider how investigation is related to problem-solving heuristics after establishing that investigation is a means to solve closed problems. The chapter ends with some implications for teaching.

2 Relationship between Problem Solving and Investigation

Whether a situation is a problem or not depends on the particular individual (Henderson & Pingry, 1953). If the person is "*unable to proceed directly* to a solution" (Lester, 1980, p. 30), then the situation is a problem to him or her. Reys, Lindquist, Lambdin, Smith, and Suydam (2004) believed that this difficulty must require "some creative effort and higher-level thinking" (p. 115) to resolve. Thus most textbook 'problems' are actually not problems to many students partly because they know how to 'solve' them and partly because the main purpose of these 'problems' is to practise students in the procedural skills that have been taught in class earlier (Moschkovich, 2002). Therefore, it may be a better idea to use the term 'mathematical task' instead of 'mathematical problem' when we are referring to the task itself. For example, the Professional Standards for Teaching Mathematics (NCTM, 1991) used the phrase 'mathematical tasks' instead of 'mathematical problems' (see, e.g., p. 25) and Schoenfeld (1985) wrote, "... being a 'problem' is not a property inherent in a *mathematical task* [emphasis mine]" (p. 74). However, we do use the terms 'mathematical problems' and 'problem solving' in this chapter, but whenever such terms are used, it implies that the task is a problem to the person because if otherwise, then there is no need to solve the task.

One of the contentious issues among educators concerns the closure or openness of mathematical problems. Henderson and Pingry (1953) believed that a problem must have a clearly defined goal, and Orton and Frobisher (1996) claimed that very few mathematics educators would classify mathematical investigations as problems because they were of the opinion that investigations must have an open and ill-defined goal. But we agree with Evans (1987) that if a student does not know what to do when faced with an investigation, then the investigation is still a problem to the student. Orton and Frobisher (1996) also observed that educators in some countries, e.g., the United States of America, would call investigations 'open problems'. But this phrase is an oxymoron if one holds on to the view that problems must be closed. Nevertheless, this suggests that many educators seem to separate mathematical problems from investigations in that the former must be closed while the latter must be open.

Others (e.g., Cai & Cifarelli, 2005; Frobisher, 1994) have suggested that investigation should involve both problem posing and problem solving. Although many educators have claimed that there are overlaps between problem solving and investigation, they still ended up separating them. For example, HMI (1985) stipulated that there is no clear distinction between problem solving and investigation but it still ended up separating problem solving as a convergent activity from investigation as a divergent activity partly because the writers believed that investigation should involve problem posing as well (Evans, 1987).

However, school teachers are often not so clear about the differences between problem solving and investigation. Some of them even feel that their students are doing some sort of investigation when solving certain types of closed problems (personal communication). For example, consider the following mathematical task which is closed:

> **Task 1: Handshakes**
> At a workshop, each of the 70 participants shakes hand once with each of the other participants. Find the total number of handshakes.

If students do not know how to solve this task, then this task is a problem to them. Some teachers believe that these students can begin by *investigating* what happens if there are fewer numbers of participants, which may help the students to solve the original problem. But there seems to be very little literature on this subject of solving a closed problem by investigation. However, a thorough search has revealed a few writings. For example, in the *synthesis* class in Bloom's taxonomy of educational objectives in the cognitive domain, Bloom, Engelhart, Furst, Hill, and Krathwohl (1956) wrote about the "ability to integrate the results of an *investigation* [emphasis mine] into an effective plan or solution to *solve a problem* [emphasis mine]". The *Curriculum and Evaluation Standards for School Mathematics* stipulated that "our ideas about problem situations and learning are reflected in the verbs we use to describe student actions (e.g., to investigate, to formulate, to find, to verify) throughout the Standards" (NCTM, 1989, p. 10), thus suggesting that the Standards do recognise investigation as a means of dealing with problem situations.

Yeo and Yeap (2009) tried to reconcile the differences between the view that mathematical investigation must be open and the view that investigation can occur when solving closed problems. The conflict appears to arise from the different uses of the same term 'investigation'. Just as Christiansen and Walther (1986) distinguished between a task and an activity, Yeo and Yeap (2009) differentiated between investigation as a task, as a process and as an activity. They called the following an open investigative task, rather than the ambiguous phrase 'mathematical investigation':

Task 2: Polite Numbers

Polite numbers are natural numbers that can be expressed as the sum of two or more consecutive natural numbers. For example,

$$9 = 2 + 3 + 4 = 4 + 5,$$
$$11 = 5 + 6,$$
$$18 = 3 + 4 + 5 + 6.$$

Investigate.

When students attempt this type of open investigative tasks, they are engaged in an activity, which is consistent with Christiansen's and Walther's (1986) definitions of a task and an activity. Yeo and Yeap (2009) called this an open investigative activity which involves both problem posing and problem solving: students need to pose their own problems to solve (Cai & Cifarelli, 2005). However, Yeo and Yeap (2009) observed that when students pose a problem to solve, they have not started investigating yet. This led them to separate investigation as a process from investigation as an activity involving an open investigative task.

An analogy is Pólya's (1957) four stages of problem solving for closed problems. During the first stage, the problem solver should try to understand the problem. But the person has not started solving the problem yet. The actual problem-solving process begins during the second stage when the person tries to devise a plan to solve the problem and it continues into the third stage when the person carries out the plan.

After solving the problem, the person should look back, which is the fourth stage. Therefore, the actual problem-solving process occurs in the second and third stages although problem solving should involve the first and fourth stages also: what the person should do before and after problem solving.

Similarly, when students attempt an open investigative task, they should first try to understand the task and then pose a problem to solve. However, this is before the actual process of investigation. After the investigation, the students should look back and pose more problems to solve. Therefore, there is a difference between the process of investigation and an open investigative activity: the former does not involve problem posing but the latter includes problem posing. From this point onwards, the term 'investigation' will be used in this chapter to refer to the process while the activity will be called an 'open investigative activity'. This distinction is important because we would like to argue that investigation can occur when solving closed problems. But first, we need to characterise what investigation is.

Yeo and Yeap (2009) observed that when students investigate during an open investigative activity, they usually start by examining specific examples or special cases which Mason, Burton, and Stacey (1985) called specialising. The purpose is to search for any underlying pattern or mathematical structure (Frobisher, 1994). Along the way, the students will formulate conjectures and test them (Bastow, Hughes, Kissane, & Mortlock, 1991). If a conjecture is proven or justified, then generalisation has occurred (Height, 1989). Thus investigation involves the four mathematical thinking processes of specialising, conjecturing, justifying and generalising, which Mason et al. (1985) applied to problem solving involving closed problems. Therefore, mathematical investigation can occur not only in open investigative activities but also in closed problem solving. But if investigation must involve problem posing, then investigation cannot happen when solving closed problems. This is why the separation of problem posing in open investigative activities from the process of investigation is very important.

Hence, if we view investigation as a process involving specialising, conjecturing, justifying and generalising, then we can solve closed mathematical problems by investigation when we are stuck.

3 Solving Mathematical Problems by Investigation

In this section, we will illustrate how investigation can help teachers and students to solve two closed mathematical problems when they are stuck, and how the result of an investigation can be used to develop a more formal or rigorous proof for their conjectures. Furthermore, we deliberate two important issues: whether induction is proof and how heuristics are related to investigation. Let us start by looking at the following task:

Task 3: Series

Find the value of $\frac{1}{1}+\frac{1}{1+2}+\frac{1}{1+2+3}+...+\frac{1}{1+2+3+...+2008}$.

This task was given to a group of in-service primary school teachers during a workshop at Mathematics Teachers Conference 2008 in Singapore. All of them had not seen this question before and they did not know how to solve it immediately, so this was a problem to them. Most of them were stuck: they did not even know how to begin. After some pondering, some of them tried to evaluate the denominators of all the fractions but it led to nowhere. So the first author guided them to investigate some specific examples by starting with smaller sums, i.e., what is the sum of the first two fractions, the sum of the first three fractions, etc., to see if there is any pattern:

$$S_2=\frac{1}{1}+\frac{1}{1+2}=\frac{4}{1+2}$$

$$S_3=\frac{1}{1}+\frac{1}{1+2}+\frac{1}{1+2+3}=\frac{9}{1+2+3}$$

$$S_4=\frac{1}{1}+\frac{1}{1+2}+\frac{1}{1+2+3}+\frac{1}{1+2+3+4}=\frac{16}{1+2+3+4}$$

$$S_5=\frac{1}{1}+\frac{1}{1+2}+\frac{1}{1+2+3}+\frac{1}{1+2+3+4}+\frac{1}{1+2+3+4+5}=\frac{25}{1+2+3+4+5}.$$

Some teachers were able to observe that $S_n = \dfrac{n^2}{1+2+3+...+n}$. The sum of the numbers in the denominator can be found easily as $\dfrac{n(n+1)}{2}$, so $S_n = \dfrac{2n}{n+1}$. Therefore, $S_{2008} = \dfrac{4016}{2009}$.

Unfortunately, most of the teachers thought that this was the answer. Some of them knew that this was only a conjecture because the observed pattern might not be true but they forgot to test the conjecture, while most of them did not even realise that this was only a conjecture. This is probably due to how they were taught number patterns in schools when they were students themselves, and now they are teaching their students the same thing: there is always a unique answer for the missing term in a sequence. For example, in the following sequence, what is the next term?

1, 4, 7, ____

Most of the teachers were taught that the answer must be 10 and so it is unique. However, the missing term is only 10 if the sequence is an arithmetic progression, in which case, the general term is $T_n = 3n - 2$. In theory, the next term can be any number. For example, the fourth term for the above sequence can be 16 if the general term is $T_n = n^3 - 6n^2 + 14n - 8$ (the reader can check that $T_1 = 1$, $T_2 = 4$, $T_3 = 7$ and $T_4 = 16$ using this formula). If you want the missing term in the above sequence to be any number, e.g., 22, all you need to do is to form and solve four simultaneous equations with four unknowns, and a polynomial with four parameters is of degree 3, i.e., the cubic polynomial $T_n = an^3 + bn^2 + cn + d$. So the four equations are:

$$\begin{aligned}
T_1 &= a+b+c+d = 1 \\
T_2 &= 8a+4b+2c+d = 4 \\
T_3 &= 27a+9b+3c+d = 7 \\
T_4 &= 64a+16b+4c+d = 22.
\end{aligned}$$

Solving the equations simultaneously, we obtain $a = 2$, $b = -12$, $c = 25$ and $d = -14$. So $T_n = 2n^3 - 12n^2 + 25n - 14$ (the reader can check that $T_1 = 1$, $T_2 = 4$, $T_3 = 7$ and $T_4 = 22$ using this formula). However, the coefficients may not always be 'nice' integral values or the simultaneous equations may have no solutions. For the latter, you can always try another polynomial that has more parameters, e.g., a polynomial of degree 4, and sooner or later, you will find a suitable polynomial. You can even try non-polynomials like a sine function.

Therefore, there is no unique answer for the missing term of a sequence. The answer that we want when we set this type of question is 'the most likely number' and what this means is that we prefer the formula for the general term to be less complicated. Thus the more terms we give for a sequence, the pattern should become more obvious and most of us may agree on one 'most likely number'. For example, 'the most likely number' for the missing term in the above sequence is 10 but some people may disagree. So, to avoid ambiguity, if we increase the number of given terms as shown below, then fewer people would disagree that 'the most likely number' for the missing term in the following sequence is 10, although it can still be any other number if we settle for a complicated formula for the general term, such as a polynomial of degree 6.

1, 4, 7, ____, 13, 16, 19

However, we cannot go for 'the most likely number' if the sequence has a context and is linked to some underlying pattern. For example, if we just consider the following sequence, then 'the most likely number' is 32 because the general term $T_n = 2^{n-1}$ is less complicated than a formula such as $T_n = {}^nC_4 + {}^{n-1}C_2 + {}^nC_1$.

1, 2, 4, 8, 16, ____

But if this sequence has a context and is linked to some underlying pattern, then we cannot just assume that the missing term is 32. For example, consider the following circle:

Figure 1. Circle with five points

There are five arbitrary points on the circumference of the circle, and each point is connected to every other point by a chord such that no three chords interest at the same point inside the circle. The chords divide the circle into regions. In this case, when $n = 5$ (where n is the number of points on the circumference of the circle), there are 16 regions inside the circle. If we consider the case when $n = 1, 2, 3, 4, 5, \ldots$, then the total number of regions inside the circle, T_n, will form the following sequence:

$$1, 2, 4, 8, 16, \ldots$$

If $n = 6$, what will be the total number of regions? The teachers in the workshop predicted that there would be 32 regions although a few of them suspected that this might not be the answer, or else the first author would not be giving them this counter example. Then the teachers counted the total number of regions for the following circle manually:

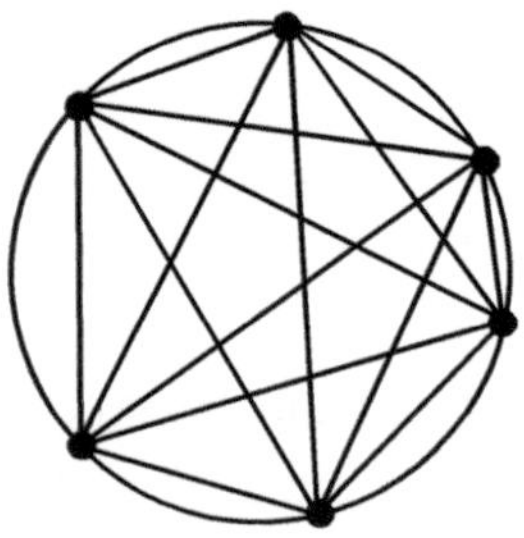

Figure 2. Circle with six points

When they found out that there were only 31 regions in the circle in Figure 2, some of them thought that they had counted wrongly and so they recounted the number of regions, while others realised that it was possible to have a sequence as follows:

$$1, 2, 4, 8, 16, 31, \ldots$$

However, some of them concluded that the above sequence has no pattern. The first author reiterated that there is still a pattern in the above sequence, but the underlying pattern is not $T_n = 2^{n-1}$ which is the 'more obvious' observed pattern in the sequence 1, 2, 4, 8, 16, … In fact, there is even a formula for the total number of regions: $T_n = {}^nC_4 + {}^{n-1}C_2 + {}^nC_1$ (the reader can check that $T_1 = 1$, $T_2 = 2$, $T_3 = 4$, $T_4 = 8$, $T_5 = 16$ and $T_6 = 31$ using this formula).

Let us return to the observed pattern in Task 3. The teachers finally realised that this was only a conjecture and they needed to test it. At first, no one was able to prove or refute it. After some time, a teacher managed to develop a rigorous proof. In fact, this teacher did not even solve the problem by investigation: she did not follow the hint of the first author above but she did the following on her own:

$$\begin{aligned}
&\frac{1}{1}+\frac{1}{1+2}+\frac{1}{1+2+3}+\frac{1}{1+2+3+4}+\ldots+\frac{1}{1+2+3+\ldots+2008} \\
&=1+\frac{1}{3}+\frac{1}{6}+\frac{1}{10}+\ldots+\frac{1}{1+2+3+\ldots+2008} \\
&=1+\left(1\div\frac{2\times 3}{2}\right)+\left(1\div\frac{3\times 4}{2}\right)+\left(1\div\frac{4\times 5}{2}\right)+\ldots+\left(1\div\frac{2008\times 2009}{2}\right) \\
&=1+\frac{2}{2\times 3}+\frac{2}{3\times 4}+\frac{2}{4\times 5}+\ldots+\frac{2}{2008\times 2009} \qquad \nwarrow \text{Line \#3} \\
&=1+\left(\frac{2}{2}-\frac{2}{3}\right)+\left(\frac{2}{3}-\frac{2}{4}\right)+\left(\frac{2}{4}-\frac{2}{5}\right)+\ldots+\left(\frac{2}{2008}-\frac{2}{2009}\right) \qquad \leftarrow \text{Line \#5} \\
&=1+\frac{2}{2}-\frac{2}{2009} \\
&=\frac{4016}{2009}
\end{aligned}$$

All the other teachers were very impressed that this teacher was able to devise such a proof[1]. The first author asked the teacher how she managed to think of Line #3 and Line #5 which were the key steps in her proof, but she herself could not explain how and why she did it this way. All the other teachers agreed that they themselves would never have thought of this type of rigorous proofs that seem to come out of nowhere, which agrees with what Lakatos (1976) wrote when he observed that "it seems impossible that anyone should ever have guessed them" (p. 142).

There is a more elegant but similar proof:

$$\text{Let } S=\frac{1}{1}+\frac{1}{1+2}+\frac{1}{1+2+3}+\frac{1}{1+2+3+4}+\ldots+\frac{1}{1+2+3+\ldots+2008}$$

$$=1+\frac{1}{3}+\frac{1}{6}+\frac{1}{10}+\ldots+\frac{1}{1+2+3+\ldots+2008}.$$

$$\text{Then } \frac{1}{2}S=\frac{1}{2}+\frac{1}{6}+\frac{1}{12}+\frac{1}{20}+\ldots+\frac{1}{2(1+2+3+\ldots+2008)}$$

$$=\frac{1}{1\times 2}+\frac{1}{2\times 3}+\frac{1}{3\times 4}+\frac{1}{4\times 5}+\ldots+\frac{1}{2008\times 2009}$$

$$=\left(1-\cancel{\frac{1}{2}}\right)+\left(\cancel{\frac{1}{2}}-\cancel{\frac{1}{3}}\right)+\left(\cancel{\frac{1}{3}}-\cancel{\frac{1}{4}}\right)+\left(\cancel{\frac{1}{4}}-\cancel{\frac{1}{5}}\right)+\ldots+\left(\cancel{\frac{1}{2008}}-\frac{1}{2009}\right)$$

Line #5

$$=1-\frac{1}{20009}$$

$$=\frac{2008}{2009}$$

$$\therefore S=\frac{4016}{2009}$$

Similarly, most people would never have thought of finding half the sum in this second proof. But how did the originator of this proof know what to do? The person most likely had to do some investigation first.

---

[1] Actually, there is more to the (first) proof than is shown here. There must be good reasons to believe that the patterns in Lines #3 and #5 will continue. We will leave it to the reader to find the reasons.

However, what might have helped in the investigation were some prior mathematical knowledge and skills which the person might have relied upon, which Schoenfeld (1985) called resources which were necessary for effective problem solving. For example, the person might have known the method of differences (which is the key step of the proof: see Line #5 of the second proof), and he or she might also be familiar with expressing $\frac{1}{n(n+1)}$ as $\frac{1}{n}-\frac{1}{n+1}$. The person might also have recalled that the numbers 1, 3, 6, 10, … , which appear in the denominators of the original series, are triangular numbers, and that the general term for triangular numbers is $T_n = \frac{1}{2}n(n+1)$, which is one step away from getting $\frac{1}{n(n+1)} = \frac{1}{n}-\frac{1}{n+1}$. These might have helped the person to think of starting with half the sum after some investigation. But if anyone does not have all these resources at his or her disposal, then the person may have to do more investigation to discover these first, or perhaps the person can conjure the first proof provided by the teacher above (this teacher has admitted that she knows the method of differences) and then refine it later to become a more elegant proof like the second one.

To summarise, this example (Task 3) illustrates the two main approaches to solve a closed mathematical problem: by investigation or by 'other means' (which is rigorous proof in this case), and that very few teachers were able to solve it using a rigorous proof directly.

Let us look at another example: the *Handshakes* task in the previous section (see Task 1). The first author has given this task to primary and secondary school students, and pre-service and in-service teachers. Some of the teachers and students have seen this question before, and they were able to give the answer almost immediately, so this task was not a problem to them. For those who saw this for the first time and were unable to solve it immediately, this was a problem to them. After a while, the teachers and the better students were able to solve it by 'other means', which in this case is simple deductive reasoning: since the first participant must shake hand with the other 69 participants, the second participant must shake hand with the remaining 68 participants and so forth, then the total number of handshakes is 69 + 68 + 67 + … + 1. Some high-ability students can even use a combinatorics argument that the total number of handshakes is equal to the total number of different pairs of participants, i.e., ${}^{70}C_2$, because every different pair of participants will give rise to one distinct handshake. This type of deductive proofs, unlike the formal proofs for Task 3, is within the grasp of many teachers and students.

But for the weaker ones who were unable to reason it in this way, many of them tried to solve the problem by drawing a diagram for smaller numbers of participants (see Figure 3 where n is the number of participants and T_n is the total number of handshakes) in order to observe some patterns so as to generalise to 70 participants. This is specialising in order to form a conjecture towards a generalisation, which are essentially the core processes in a mathematical investigation.

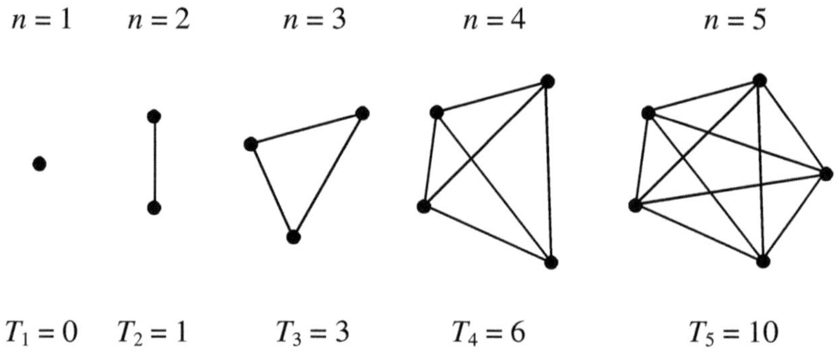

Figure 3. Handshakes task

Many of them were able to observe from their diagrams that the total number of handshakes for n participants is 0, 1, 3, 6, 10, … for n = 1, 2, 3, 4, 5, … respectively. However, most of them were unable to find a formula for the general term of this sequence. But they were able to observe this pattern:

$$0, \quad 1, \quad 3, \quad 6, \quad 10, \quad \ldots$$
$$+1 \quad +2 \quad +3 \quad +4$$

Using this pattern as a scaffold, the first author guided the teachers and students by asking them how to obtain T_4 from T_2. This enabled most of them to observe that $T_4 = 1 + 2 + 3$. Similarly, to obtain T_5 from T_2, most of the teachers and students were able to see that $T_4 = 1 + 2 + 3 + 4$. Therefore, they were able to observe that $T_{70} = 1 + 2 + 3 + \ldots + 69$, which is the total number of handshakes for 70 participants. Unfortunately, most of them, including the teachers and the better students, thought that this was the answer, without realising that this was only a conjecture to be proven or refuted. If the conjecture is wrong, you can refute it by using a counter example. But if the conjecture is correct, then do you really need a formal or rigorous proof to prove it? Some educators (e.g., Holding, 1991; Tall, 1991) believe in using rigorous proofs while others (e.g., Mason et al., 1985) support justification using the underlying mathematical structure. We shall illustrate these two approaches of justification using the *Handshakes* task.

The first author began by asking the teachers and students whether there was any reason to believe that the observed pattern would continue. Not a single person was able to find a reason. So the first author guided them with this question: if you go from $T_4 = 1 + 2 + 3$ to T_5, what happens? Some of them were able to observe that if you add the fifth participant to T_4, then the fifth participant must shake hand with each of the four participants, so there are four additional handshakes and thus $T_5 = 1 + 2 + 3 + 4$. Using the same argument, if you add the sixth participant to T_5, then the sixth participant must shake hand with each of the five participants, so there are five additional handshakes and thus $T_6 = 1 + 2 + 3 + 4 + 5$. Therefore, this is a good reason to believe that the observed pattern will continue in this manner because this argument can always be applied from T_n to T_{n+1}. But this is not a proof. However, Mason et al. (1985) believed that this type of argument using the underlying mathematical structure is good enough for school students.

The next question is how to guide these teachers and students to construct a more rigorous proof for their conjecture from their investigation. From the underlying mathematical structure discovered in the above investigation (i.e., if you add one participant to n participants, then the new participant must shake hand with the n participants, thus resulting in n additional handshakes and so the total number of handshakes for T_{n+1} is $1 + 2 + 3 + \ldots + n$), a few teachers and students were able to realise that they could use the same argument in the reverse manner: start from the first participant, and he or she has to shake hands with all the other 69 participants; then the second participant has only 68 participants to shake hand with, and so forth; thus the total number of handshakes for T_{70} is $69 + 68 + 67 + \ldots + 1$. In this way, the teachers and students have managed to use their investigation to develop a more rigorous proof for their conjecture. This agrees with what Pólya (1957) believed when he wrote that "we need heuristic reasoning when we construct a strict proof as we need scaffolding when we erect a building" (p. 113). According to Pólya, heuristic reasoning is based on induction or analogy, but both induction and analogy involve specialising in order to discover the underlying mathematical structure. Therefore, Pólya's idea of heuristic reasoning is very similar to the concept of the process of investigation outlined in the previous section.

One major issue to deliberate in this section is whether induction is proof. Yeo and Yeap (2009) believed that the problem lies in the different meanings of the terms 'induction', 'inductive observation' and 'inductive reasoning'. If students observe a pattern when specialising, the pattern is only a conjecture and Lampert (1990) called this 'inductive observation'. But if students use the underlying mathematical structure (Mason et al., 1985) to argue that the observed pattern will always continue, then it involves rather rigorous reasoning and so this can be called 'inductive reasoning' (Yeo & Yeap, 2009). Thus there is a big difference between inductive observation and inductive reasoning: inductive observation is definitely not a proof but inductive reasoning is considered a proof by some educators (e.g., Mason et al., 1985). Unfortunately, some educators (e.g., Holding, 1991) have used the phrase 'inductive reasoning' to mean 'inductive observation'. The same goes for the word 'induction': it can mean either 'inductive observation' or 'inductive reasoning' or both. For example, Pólya's (1957) idea of induction is inductive observation only. Therefore, whether induction is proof or not depends on which meaning you attach to the term 'induction'. In this chapter, the term 'induction' is used to include both inductive observation and inductive reasoning.

Another main issue to discuss in this section is the relationship between heuristics and investigation as a means to solve closed mathematical problems. Literature abounds with problem-solving heuristics (see, e.g., Pólya, 1957; Schoenfeld, 1985) but very few of them mention the use of investigation to solve closed problems, probably because few educators have ever characterised the process of investigation. Now that we have observed that investigation involves the four core processes of specialising, conjecturing, justifying and generalising, we can compare investigation with heuristics. Any heuristic that makes use of specialising can be considered an investigation (Yeo & Yeap, 2009). For example, if students use the heuristic of systematic listing or the heuristic of drawing a diagram for some specific cases, then it involves specialising and so this can be viewed as an investigation from another perspective. But if students use a deductive argument *directly*, then this is not an investigation. It does not mean that students cannot use deductive reasoning during an investigation. For example, students can use a deductive argument when proving a conjecture that is formulated during their investigation.

4 Conclusion and Implications

Differentiating between investigation as a task, as a process and as an activity has helped to separate problem posing from the process of investigation. This is important because if investigation entails both problem posing and problem solving, then investigation cannot happen during problem solving. Characterising the process of investigation as involving specialising, conjecturing, justifying and generalising, it becomes clear that investigation can also occur when solving closed mathematical problems. This agrees with what some teachers believe when they ask their students to investigate to solve a closed problem but most of them have no idea what investigation actually involves. If teachers have a vague idea of what investigation entails, then they may not be able to teach their students how to investigate properly (Frobisher, 1994). Therefore, the implication of defining the process of investigation more clearly in this chapter is to help teachers understand more fully what investigation means and how to help their students to investigate more effectively by focusing on each of the core thinking processes of specialising, conjecturing, justifying and generalising.

Another implication for teaching is how to make use of the results of an investigation as a scaffold to construct a more rigorous proof for a conjecture (Pólya, 1957) instead of conjuring a formal proof out of nowhere (Lakatos, 1976).

References

Bastow, B., Hughes, J., Kissane, B., & Mortlock, R. (1991). *40 mathematical investigations* (2nd ed.). Western Australia: Mathematical Association of Western Australia.

Bloom, B. S., Engelhart, M. D., Furst, E. J., Hill, W. H., & Krathwohl, D. R. (1956). *Taxonomy of educational objectives: The classification of educational goals. Handbook I. Cognitive domain.* New York: David McKay.

Cai, J., & Cifarelli, V. (2005). Exploring mathematical exploration: How two college students formulated and solve their own mathematical problems. *Focus on Learning Problems in Mathematics, 27*(3), 43-72.

Christiansen, B., & Walther, G. (1986). Task and activity. In B. Christiansen, A. G. Howson, & M. Otte (Eds.), *Perspectives on mathematics education: Papers submitted by members of the Bacomet Group* (pp. 243-307). Dordrecht, The Netherlands: Reidel.

Evans, J. (1987). Investigations: The state of the art. *Mathematics in School, 16*(1), 27-30.

Frobisher, L. (1994). Problems, investigations and an investigative approach. In A. Orton & G. Wain (Eds.), *Issues in teaching mathematics* (pp. 150-173). London: Cassell.

Height, T. P. (1989). *Mathematical investigations in the classroom.* Australia: Longman Cheshire Pty.

Henderson, K. B., & Pingry, R. E. (1953). Problem solving in mathematics. In H. F. Fehr (Ed.), *The learning of mathematics: Its theory and practice* (pp. 228-270). Washington, DC: National Council of Teachers of Mathematics.

HMI. (1985). *Mathematics from 5 to 16.* London: Her Majesty's Stationery Office (HMSO).

Holding, J. (1991). *The investigations book: A resource book for teachers of mathematics.* Cambridge, UK: Cambridge University Press.

Lakatos, I. (1976). *Proofs and refutations: The logic of mathematical discovery.* New York: Cambridge University Press.

Lampert, M. (1990). When the problem is not the question and the solution is not the answer: Mathematical knowing and teaching. *American Educational Research Journal, 27,* 29-63.

Lee, M., & Miller, M. (1997). *Real-life math investigations: 30 activities that help students apply mathematical thinking to real-life situations.* New York: Scholastic Professional Books.

Lester, F. K., Jr. (1980). Problem solving: Is it a problem? In M. M. Lindquist (Ed.), *Selected issues in mathematics education* (pp. 29-45). Berkeley, CA: McCutchan.

Mason, J., Burton, L., & Stacey, K. (1985). *Thinking mathematically* (Rev. ed.). Wokingham, UK: Addison-Wesley.

Moschkovich, J. N. (2002). An introduction to examining everyday and academic mathematical practices. In E. Yackel (Series Ed.) & M. E. Brenner & J. N. Moschkovich (Monograph Eds.), *Everyday and academic mathematics in the classroom* (pp. 1-11). Reston, VA: National Council of Teachers of Mathematics.

National Council of Teachers of Mathematics. (1989). *Curriculum and evaluation standards for school mathematics.* Reston, VA: National Council of Teachers of Mathematics.

National Council of Teachers of Mathematics. (1991). *Professional standards for teaching mathematics.* Reston, VA: National Council of Teachers of Mathematics.

Orton, A., & Frobisher, L. (1996). *Insights into teaching mathematics.* London: Cassell.

Pirie, S. (1987). *Mathematical investigations in your classroom: A guide for teachers.* Basingstoke, UK: Macmillan.

Pólya, G. (1957). *How to solve it: A new aspect of mathematical method* (2nd ed.). Princeton, NJ: Princeton University Press.

Reys, R. E., Lindquist, M. M., Lambdin, D. V., Smith, N. L., & Suydam, M. N. (2004). *Helping children learn mathematics* (7th ed.). Hoboken, NJ: John Wiley & Sons.

Schoenfeld, A. H. (1985). *Mathematical problem solving.* Orlando, FL: Academic Press.

Tall, D. (1991). *Advanced mathematical thinking.* Dordrecht, The Netherlands: Kluwer Academic Press.

Yeo, J. B. W., & Yeap, B. H. (2009). *Mathematical investigation: Task, process and activity* (Tech. Rep. ME2009-01). National Institute of Education, Nanyang Technological University, Singapore.

Chapter 8

Generative Activities in Singapore (GenSing): Pedagogy and Practice in Mathematics Classrooms

Sarah M. DAVIS

This chapter discusses a new technology-supported classroom pedagogy, Generative Activities. These activities are rooted in the tradition of function-based algebra and utilize a classroom network of handheld devices. A curricular intervention was done where the algebra topics in the Secondary 1 Scheme of Work were rearranged into three structural concepts; equals (where two expressions are everywhere the same), equivalence (the intersection of two expressions) and concepts of the linear functions (slope, rate, intercept). Activities were created using the Generative Design principles of space creating play, dynamic structure, agency and participation. The teachers involved with the project put in much hard work on changing their pedagogical practices to encourage creativity and take advantage of the classroom network to further students' conceptual understanding of mathematics. Results from the Singapore classrooms show the creative potential of these types of activities.

1 Introduction

To function in the 21st century students need deep conceptual understanding of mathematics, specifically algebra as it is the gatekeeper to higher-level mathematics. Research in mathematics education has shown that educational environments in which students engage in mathematics as mathematicians, sharing and discussing ideas, fosters this deeper understanding (Boaler, 1998, 2002; Lampert, 1990, 2001). An approach that shifts the focus from procedural computation with its emphasis on right and wrong to an emphasis on ideas and structure, requires a change in the pedagogical practices of the classroom (Lampert, 1990). Traditional algebraic instruction has students arriving at Secondary 3 and 4 with only a computation understanding of algebraic topics. It has created students without robust conceptual connections between the different algebraic representations of graph, expression and table. Representational fluency, the ability to switch between representations of mathematical concepts and to fully understand those representations, is believed to be crucial to students success in mathematics (National Council of Teachers of Mathematics, 2000).

> Representations should be treated as essential elements in supporting students' understanding of mathematical concepts and relationships; in communicating mathematical approaches, arguments, and understandings to one's self and to others; in recognizing connections among related mathematical concepts; and in applying mathematics to realistic problem situations through modeling. (National Council of Teachers of Mathematics, 2000, pg. 66)

Successful models of instruction in algebra have employed a focus on multiple representations and Generative Design principles as the basis of instruction (Kaput, 1995, 1998; Stroup, Ares, Hurford, & Lesh, 2007; Stroup & Davis, 2005; Stroup, Kaput, Ares, Wilensky, Hegedus, Roschelle, Mack, Davis, Hurford, 2002). While the generative activities discussed in the chapter may not be what one expects to see when hearing the phrase "problem solving", we believe that these activities problematize mathematics. They create opportunities for students to solve mathematical problems as a class, to harness the ideas of everyone in the room to create a host of artifacts to explore. The exploration of these artifacts, or representations of algebraic concepts, builds representational fluency. This fluency, the ability to navigate different ways of representing mathematical phenomena (e.g., symbolic, graphical, tabular, verbally and with gesture), facilitates the development of problem solving skills.

The projects discussed in this chapter, GenSing Pilot and GenSing1, have implemented generative activities in mathematics classrooms in Singapore (Davis, 2007). Both projects point to the effectiveness of a focus on powerful mathematical ideas and new styles of interaction. This chapter will discuss the theoretical foundations of Generative Design with examples of classroom work from Singapore to illuminate how the theory relates to practice.

2 What is a Generative Activity?

At the core of Generative Design is the belief that a well-designed activity should "never ask a question with only one right answer" (Judah Shwartz, Harvard). One of the great shortcomings of traditional instruction is that it teaches students that "*doing* mathematics means following the rules laid down by the teacher; *knowing* mathematics means remembering and applying the correct rule when the teacher asks a question" (Lampert, 1990, p 32). Generative activity work to change that perception, instead of having the students simplify $\frac{25x + 7x}{16}$, they are challenged to come up with 3 functions the same as 2x. In this way *doing mathematics* is a process involving experimentation, creativity and even failure (finding out what doesn't work can be as valuable as finding out what does), and *knowing mathematics* is using the underlying structural concepts as to find answers that meet the criteria (in this case that two expressions are equivalent if their graphs are the same).

Before we continue with the theoretical aspects of Generative Design is critical for the reader to have a good understanding of what Generative Activities are, what they look like in a classroom and how they contribute to a deep understanding of mathematics. To facilitate the reader's understanding of the types of activities and pedagogy this paper will be discussing, the following narrative of an activity being done at a Singapore Secondary school given below.

3 Narrative

Function Activity 2, the class begins: The students log into the TI Navigator network and are shown a Cartesian plane from -10 to10 on the X axis and -20 to 20 on the Y axis. On their TI 84+ calculator they see one point (their point), which they can move around their screen using the arrow keys on the device. In the upfront projection of the teacher computer, all of the students' points are visible. The teacher computer updates in real time; as the students move their point on their calculator, their icon in the group display also moves. The calculator is their private-space, the projection of the teacher computer is the public-space (see Figure 1).

As with any new manipulative in the classroom, students need to be given an opportunity to explore prior to settling in to the core task. The students are given a series of "playful" tasks to help them learn the point submission interface and explore the possibilities: "Move to a place on the graph where your X and Y coordinates are positive. Move to the second quadrant. Move to the origin." This allows the students to learn the calculator interface and get familiar with the group representation before they are asked to do a more robust mathematical task. As will be mentioned in other places in the paper, the students are not playing (as in a game with no instructional goals), they are playfully interacting with mathematics. While they are racing each other to get to the next designated location, they are learning the interfaces and reviewing concepts of the coordinate plane. The students are now ready to use the technology as a tool to gain mathematical knowledge.

"Find a point whose Y value is twice the X value", the teacher challenges the class. This first activity has the students embodying the definition of function. No two students can have the same X value but different Y values. They are free to talk with their peers about what this means, but each student must submit their own point. For the first minutes of this task, the upfront display shows only the coordinate plane with the points moving to follow the rule. After the students have had a chance to try and figure out what "make your Y twice your X" means, the teacher changes the upfront display so that in addition to the graph window, the students can now see a list of the coordinate pairs created by each point. For students who are struggling to translate the verbal description to a mathematical relation, seeing the numeric values of their classmates' points can help to scaffold their understanding. The "more knowledgeable peer" (Vygotsky, 1978) does not need to be sitting next to them, the group display provides a venue for all students to mentor and be mentored interchangeably. With this just-in-time modification of the task, the teacher keeps all students engaged at a level and in a way that is meaningful for them. After another minute, the activity is stopped. This freezes the student devices and allows for the submitted data to be discussed.

Figure 1. The group public-space is shown on the left and the students' private-space is shown on the right

"Do you see a pattern in the data? What is the pattern? Do you see any points that don't seem to fit the pattern? Can we fix those points?" The teacher and students explore the data. The class works together to fix all of the points that were not following the rule. The activity is re-started in function mode. Now, instead of controlling a point, the students can submit functions. On their calculator screen they can see the set of points created by the class and have a prompt to input a function to fit the data.

As with the points, while students are first struggling to figure out what it would mean symbolically for Y to be twice X, the upfront display only shows the set of points and any graphs of submitted functions. After a number of students have found a function to fit the data set, the teacher changes the display to show both the graph window and the equation window. Once the students have found one function to go through the data set, they are challenged to find two additional functions that also go through the data set. Displaying the graphical and symbolic forms of the functions accomplishes two things. First, for students who are still struggling to find a first function to model the data set, seeing the work of their peers can help them figure out a correct equation. More importantly, the display of submitted functions helps to build a sense of theater among the students. Wanting their function to standout, to get noticed, students start to get more and more creative with the functions they submit (Davis, 2003). For a sample of student created functions, see Table 1.

The rich data in the group display gives the teacher formative insight as to which concepts the students are or are not comfortable (for example if there are no decimal coefficients). With this knowledge, the teacher can guide discussion by privileging certain functions in the group display. "Oh look, there is one using division of variables. Wow, this one uses 15 terms. Can anyone figure out an expression using multiplication with negative numbers?" As these different functions are remarked upon by the teacher, or noticed by peers, students are motivated to ascertain how they were created, and then try to submit one even more interesting. Armed with the structural knowledge that two expressions are the same if their graph is the same, students can experiment and discover rules for what works. If the graphs are the same, they got it right.

The teacher and class then discus the submitted functions, admiring and analyzing the ones that were correct, and working together they correct the ones that had errors. As the students focus on an incorrect function, they have to try and determine what the person who submitted it was trying to accomplish, and then correct the mistake. Was the error in order of operations? Did the person make an addition error? Is there an error due to a miss match in factoring variables? In fixing functions submitted by their classmates, students focus on the strategies used by other people.

4 GenSing Theoretical Foundations

There is a tendency to view schooling and especially mathematics as individual work done in a group setting. Generative Activities, facilitated by classroom networks, allow for the classroom to become a true group space, a place where all individuals interact to form rich and unique digital artifacts. These group artifacts represent a collective intelligence that the teacher and students can investigate to come to deeper understandings of mathematical concepts. By combining these powerful networks with the new pedagogical practices and curricular goals of Generative Design, classroom environments can be created that thrive on the variety of answers students can create. This environment allows the focus to be placed on ideas, and the exploration of many different answers makes powerful mathematical discourses possible.

5 Generative Design Theory and Practice

Such Generative Activities as described above are at the heart of the GenSing projects. This section will weave together the design principles upon which Generative Activities are built and examples of what those theoretical principles provoke in actual practice.

In 1995, Jim Kaput outlined what he felt would need to happen to make algebra more accessible to more students and how that change would most likely occur. First he laid out three dimensions of reform for algebra, *breadth*, *integration* and *pedagogy* (Kaput, 1995). To achieve breadth one must interweave the many different facets of what it is to do algebra; modeling, working with functions, generalization and abstraction. In addition to breadth within mathematics, he felt it would be important for algebra to be integrated across other subjects. Finally, he stated that the pedagogy for teaching algebra, especially as supported by new technologies would have to change. Kaput then went on to outline three phases of reform. Near term, where existing curriculums were enhanced by the use of new technologies, mid-term where algebra was more significantly implemented in the middle grades and long term where the mathematics curriculum would be totally restructured across all grade levels and algebra as a specific course would disappear.

This section will discuss the ways in which the GenSing project has used Kaput's vision of reform to shape classroom implementation in Singapore. Specifically the theories and practices surrounding breadth, pedagogy and near and mid-term reform.

5.1 *Breadth*

The GenSing intervention uses a curriculum of function-based algebra supported by a classroom network (TI Navigator), and is grounded by the belief that most algebraic topics can fit within three key areas; "*equivalence* (of functions), *equals* (one kind of comparison of functions), and a systematic engagement with the *linear function*" (Stroup, Carmona, & Davis, 2005, p 3). The classroom integration of the TI Navigator network were described in the narrative. This section will focus on large structural ideas on which the curriculum was built.

Within typical introductory algebra topics there are three big areas of instruction; ideas of equivalence, ideas of equals and ideas of the linear functions. Equivalence is the idea that you can have two equations that look completely different in their algebraic form but graphically they are the same (Figure 2). Equivalence encompasses (but is not limited to) simplifying, factoring, combining like terms and expanding polynomials.

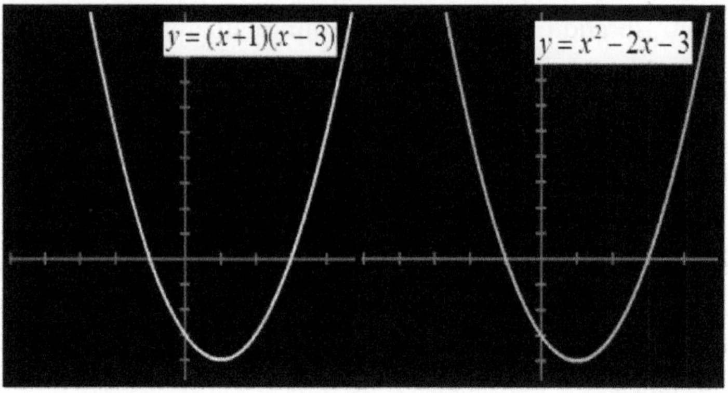

Figure 2. Equivalence: Two expressions that look nothing alike create the same graph

These ideas are usually completely disjoint for students for most of their mathematic career. Other than exactly following the rules they don't understand why what they are doing works, why it is correct. The big idea of equivalence gives students the ability to see that for expressions, there are different ways of writing the same thing.

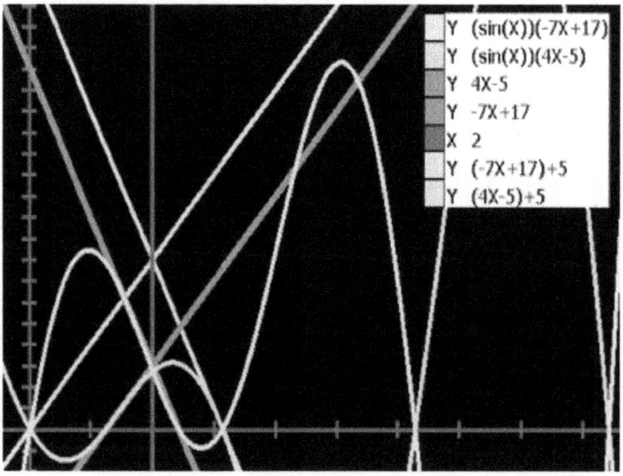

Figure 3. Equals: A special relationship between two expressions where they intersect. "Doing the same thing to both sides", preserves the solution set. Even in the extreme example of multiplying both sides by sin(x), the solution set is preserved (new solutions are added, but the original intersection point is preserved)

Equals and or Inequalities are the second big part of introductory algebra. For example, solving systems of linear equations, solving inequalities and finding where one expression is greater than another. Here the focus is not on expressions that are everywhere the same. Here the focus is on the place (or places) where two different expressions are the same; the intersection(s) (Figure 3). Taking a function-based look at the concept of equals helps students answer; Why am I supposed to do the same thing to both sides?, Why do I flip the inequality if I multiply or divide by a negative? and What does it matter if the X and the Y are different sides of the equals sign? Take for example the two expressions in Figure 4, 2x and -.5x+2. If we focus on where 2x is greater than -.5x+2, we are looking at points that are in the first quadrant. If both sides are multiplied by -1, the X-values which had been greater than, are now less than. The rule of flipping the inequality when you multiply by a negative ceases to be an obscure rule that is just memorized. The graphs literally move, the region of greater-than less-than has to be changed because the graphs aren't in the same relationship to each other any more. Without a visual representation of why the rules they are using to solve algebraic problems work, students easily confound the rules for Equals and Equivalence.

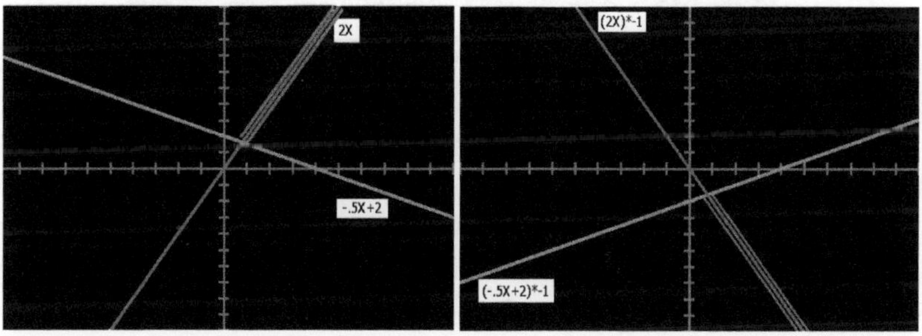

Figure 4. Graphs of Y=2X and Y=-.5X+2 and the same graphs multiplied by -1

The final idea is of linear function, specifically doing activities to separate for the student the ideas of intercept and slope. The project starts by tying ideas of physical motion to slope. Much work is done with motion detectors. This allows students to see that moving towards the motion detector is a positive slope, moving away is a negative slope, faster motion is steeper, and slower is flatter. We believe that starting with a linear function to explore ideas of slope and rate is like starting with black to explore colors. There is not enough variety or richness of information available for the students to build understanding. It is in the slowing down and the speeding up, its in the changing points in the graph that you start to make sense of where the fast parts and slow parts are. If it is always moving the same speed, if it is always a straight line, there is no complexity in it to see the changes in speed to make sense of it. For this reason, we start with messy graphs and qualitative verbal descriptions of motion.

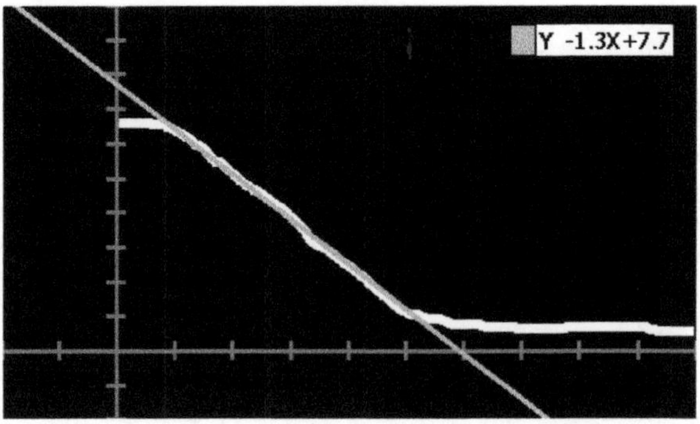

Figure 5. Motion detector graph and expression

After the students have become proficient with describing changing rates, we look at constant rates. As an example of an activity, one student will act out a rule in front of a motion detector, start at a point away from the motion detector and move at a constant moderate pace towards it. The network is used to collect the one graph and send it out to all the calculators in the classroom. The students then fit a function on top of it (Figure 5). A series of these rules are done to tie the ideas of what the person acted out to features of the expression. We start out with wiggly graphs to give students a rich environment to explore the faster and slower parts, then we move to constant motion connecting up with functions to model that motion. In this way the student is mathematizing the motion and quantifying the slope and intercept.

5.2 *Pedagogy*

The new pedagogy that the GenSing project employs is that of Generative Design. There are four key principles for designing for the generative space. Activities should have a dynamic structure, open up a space for mathematical play, allow greater agency for the students and increase participation (Stroup, Ares, & Hurford, 2005; Stroup et al., 2002).

6 Space Creating Play and Dynamic Structure

Space creating play refers to the way in which activities are structured to allow for many valid ways of participation. Play which can be seen as a negative concept in schools does not have to be perceived in that fashion. Games that students play have guidelines for appropriate participation. During generative play, within the classroom, activities also have rules and guidelines that all participants agree to abide by. The artifacts of that participation are displayed back to the group in a way that is meaningful for use in concept development. In this way, activities can be structured to allow for exploration of the mathematical space or scientific. By focusing on space creating play, activities can be designed to have a dynamic type of input where students follow rules and create a multitude of responses.

Dynamic structure referrers to the impact the emerging artifacts have on how the activity will proceed. This fits in very closely with the concept of space creating play. Space creating play is task dependent and dynamic structure is people dependent. In the $\frac{25x+7x}{16}$ example, the student response space is small. There is only one right answer. This type of question gives the student little if no ability to impact the direction of instruction. In contrast, the list of responses to the 2x question will influence how the class proceeds. The displayed responses dynamically structure what is available for discussion during the lesson. This can create some uncertainty for the teacher. There is no way to predict from one enactment to another exactly where the lesson is going that day. In one class period there might be examples of the distributive property, in another class there might be errors with multiplication of negatives to be discussed. The artifacts are a reflection of the students' ideas and current understanding. As such they can be used to do on-the-fly formative instructions. For example, if a class has just finished a unit on negative numbers and no one uses negative numbers in the functions that are being submitted, this can indicate that the students are not comfortable with negatives yet. The teacher has formative information as to what the students have and have not incorporated into their own schemes of problem solving.

7 Agency and Participation

Agency refers to the students' identity in the class, how they feel that identity is valued, and how much influence they believe they have on the content of the class. In generative activities, the entire space of responses, the basis for all classroom dialogue, is from the students. It is the ownership of the very authorship of classroom content which increases students' agency in generative activities. Anonymity of response in the display space gives students the option of expanding their agency to play different roles. In the class discussion a student can comment on an answer as if it was theirs or as if it belonged to someone else. Depending one the answer they choose to discuss, this allows them to play the role of someone who got the answer right or to hypothesize on the reasoning of someone who got it wrong. Both roles can be assumed independent of the correctness of the actual answer submitted by the student. By virtue of the anonymity in the display space all answers become everyone's answers.

This increase in agency provides opportunity for increased participation. According to Seymour Papert, a technology should have a high ceiling and a low threshold. The same can be said of generative activities. By asking questions that have more than one right answer, students are invited to participate in a way, and at a level, which is meaningful to them. Returning to the example of the 2x activity, for some students, valid participation is simply returning the original function of 2x. For others it will be exploring rational expressions and sending in 100x/50. Both are valid correct responses and let the student enter the "game" in ways that make sense to them.

Finally, participation, unlike tutoring where only one student can answer the question, every student can answer every question in the generative classroom. All responses are learning opportunities, they are either exemplars of the mathematic properties or in the case of incorrect responses they are a chance for the class to explore what the other person was thinking. What their logic process was. Fostering empathy. In generative classrooms, when teachers are reviewing responses with the class, they are encouraged to ask , "What is right about this?". What is right, then what needs to be fixed? Encouraging the students to focus on the structure of the mathematics to determine where the solution went wrong. Many wrong answers have been well thought out; just somewhere in the process there was a mathematical error.

This section has explored the foundational ideas for generative design. These ideas are what shape the school-based work of the GenSing project. The next section will share some examples of work done in Singapore.

8 Examples from Singapore Classrooms

In 2007 (GenSing Pilot) an intervention was done as a series of three, researcher lead, activities with 183 Secondary 1 students at an upper performing secondary school in Singapore. Each class was visited once every two weeks. The data was collected during the second activity (the Function Activity 2 described above). Class sizes ranged from 38-42 students. The schools' curriculum specialist and I collaborated to reorganize the Secondary 1 scheme of work and gather all of the algebraic topics covered across the year into two cohesive groupings, one lasting eight weeks and one lasting four weeks. These two groupings were then organized such that they aligned with the big three conceptual ideas. During the first eight weeks segment, concepts of equivalence, rate and linear functions were covered. During the four weeks segment, concepts of equals and inequalities were explored. In this way the entire intervention was grounded in the Generative Design literature. Additionally the topics of instruction were grouped into concepts (i.e. equivalence) and skills (i.e. factoring), where concepts were taught using generative activities and skills were taught via traditional instruction (Lesh & Doerr, 2003).

8.1 *Student learning: Powerful means of learning mathematics*

Equivalence is the major structural concept in Function Activity 2 (the activity from the narrative). The following will give examples of student data from this activity and discuss work from the GenSing Pilot and GenSing1.

Student work from both implementations showed that generative activities are surfacing important mathematical concepts; concepts that are traditionally taught in discreet chunks of memorized material. This data was created by either the rule "Make your Y twice your X" or "Make you Y 6 more than X." Using examples from Table 1, students created functions using; the distributive property Y=10(X+5)-10(5)-8X; generalizations on mathematical objects Y= PI(X+6)/PI; order of operations Y=(X^2)/X*((X^2)^2)/X/X/X/X+6; identity property of multiplication Y=100000000000X/100000000000+6; combining like terms Y=2(0.2X+0.2X+0.2X+0.2X+0.2X) and many others.

Students displayed a number of interesting strategies within their equivalent expressions. These included: Addition / Subtraction / Multiplication / Division of numbers; Addition / Subtraction / Multiplication of variables; Expressions in Simplest Form; Rational expressions; Rational 1 expressions—where the students used the concept that the numerator and denominator would cancel each other out and form a 1 (example Y=7878787878X/7878787878+6); Additive 0—these items were coded as distinct because the students used "chunks" of terms (X-X or 6-6) that totaled zero to create functions the same as 2X or X+6 (example Y=3X-X+X-X+X-X+X-X); Multiplicative 0—where the student would put in a term or a parenthetical group of terms and then multiply it by 0 to make it disappear.

Table 1

Social Strategies Student Examples

Big Numbers		**Humorous**	
Y2X	123456789X-123456787X	Y2X	2.1X-0.1X
	X10000000000-9999999998X	YX+6	98652265256X*0+X+6
	1000000X/500000	**Unique Strategy**	
	15153X-15151X	Y2X	(85X-5X)/(50-10)
	800000X/400000		(sqrt(100)-8)*X
YX+6	X+123456789-123456783		10(X+5)-10(5)-8X
	(999999999*9999999)*0+X+6		2(4X-3X)
	100000000000X/100000000000+6	YX+6	X+sqrt(36)
	20202020-20202014+X		(-2--8)+X
	(123456789-123456783)+X		PI(X+6)/PI
			X+(5^2)-19
Many Terms			
Y2X	(X*X*X*X*X*X*0)+2X		
	(X+X+X)/3*2/2*2		
	2(0.2X+0.2X+0.2X+0.2X+0.2X)		
	3X-X+X-X+X-X+X-X+X-X+X-X+X-X+X-X+X-X+X-X+X-X+X-X+X-X+X-X+X-X+X-X+X-X		
	X*X/X*X/X*X/X*X/X*2		
YX+6	XX-XX+X-X+XX-XX+6+X		
	(X^2)/X*((X^2)^2)/X/X/X/X+6		
	2X/2+58-52+3-3-9780+9780*645/645		
	X*12/6*4/4/2+6		
	X+X-X+X-X+X-X+6-6+6-6+6-6+6-6+6-6+6-6+6-6+6-6+6-6+6-6+6-6+6-6+6-6+6-6+6-6+6-6+6-6+6		

Manipulations of real numbers and non-rational algebraic expressions were the most frequent mathematical strategies adopted. None of these strategies were directly taught; they emerged from the students experimentation in the private space of the calculator. The students had previously learned that if they created two functions that made the same graph, the functions were the same. This gave them the ability to try out different combinations of terms to find things that work.

Being a mathematician is not only about math content it is also about working in a community and communicating ideas (Lampert, 1990, 2001). For this reason, mathematical strategies were not the only ones that were identified. The social strategies of the students show the beginnings of thinking of mathematics as social. Here is evidence from the GenSing Pilot. Four Social Strategies immerged from the student work: 1) Big Numbers, 2) Many Terms, 3) Unique Strategy, and 4) Humorous. These were considered to be social strategies because they were noticed by others in the classroom. Big numbers were any that had a term with five or more digits. Many Terms were functions that had five or more terms. The Unique Strategies used mathematical notions that were rare across the classes and Humorous were functions that seemed to have a tongue in cheek feel to them (for example taking a very large number and then multiplying it by 0). A vein for future research is to explore if the student's desired audience was peer or teacher attention, or if they simply wanted to be different for themselves. The act of submitting their functions to the public space, made it open for interpretation. It is in the social strategies that the importance of a space for mathematical play shines through. As in sports or other games, students explore the possibilities. They find ways to stand out, to perform. In these activities it is the Vygotskyian sense of play that is being focused on. Not an anything goes environment, but one where rules bound play and children can explore new social structures (Vygotsky, 1978). Similar to activities outside the classroom such as sports, when students see a great move, they want to copy it, or out do it.

8.2 *Teacher pedagogy: Model of teacher professional development*

The activity in the narrative above is based on a pedagogy where the teacher facilitates discussion and gets the students to explore, understand and extend their work and that of their classmates. Clearly this type of mathematics classroom requires a pedagogy different from that of lecture and independent practice. Getting teachers to change classroom practice is a challenge (Ball & Bass, 2000). Most have experienced mathematics in schooling (with great success) as a set of procedures to work through to get the correct result. There is research which links lack of conceptual understanding with a tendency to focus on procedural aspects of functions (Stein, Baxter, & Leinhardt, 1990). For these reason, the GenSing projects created a model of teacher professional development that was layered and prolonged. The first layer was teachers watching the pilot implementation where the visiting researcher worked with their students modeling the activities and pedagogical approaches. The next layer was having the teachers experience activities as students, sharing ideas and building a community of practice. The next layer was intensive technical training on how to run the equipment. The final layer was ongoing site visits, staff meetings and debrief sessions with the teachers as they were implementing the new sequence of instruction. Additionally, detailed curricular materials were created to assist the teachers in situ with orchestrating these more dynamic and interactive lessons. All activities were done with the multiple goals of increasing pedagogical content and domain knowledge and changing classroom practice (Lloyd & Wilson, 1998; Swafford, Jones, & Thronton, 1997).

8.3 *Technology innovation*

The initial data analysis done for the GenSing Pilot study was exciting and surfaced a number of interesting research questions. How do noteworthy exprcssions in the group space affect other students? How can the mass amounts of data be meaningfully organized so the teacher can reflect on student progress outside of class? What data needs to be easily accessible so that it can be re-used in another activity or additional practice?

To start the process of answering the questions, the GenSing1 project needed to create new software. First, we needed to be able to collect more than just end state data from the activities. So a script was created for continuous data collection. The new script collected all function submissions by the students with a time stamp. Additionally it also captured any changes to the data made by the teacher on the up front computer. Second, using NetLogo (Wilensky, 1999), a set of analysis tools, the GenSing Graphical Viewer (Davis & Brady, 2008a) and the GenSing Timeline Viewer (Davis & Brady, 2008b), were created to help visualize the student data. We decided to use NetLogo for it's low overhead, flexibility and modeling capability.

The first tool, GenSing Graphical Viewer imports all of the student submitted functions from a session. It is able to display all of the functions submitted for a given activity, evaluate them and write information back to the data file about the functions. The second tool is the GenSing Timeline Viewer. This software can also import the class created data file and creates a series of new views onto the data. In the first view, the Timeline Viewer (Figure 6) gives an overall idea of when activity is happening in the class; on the Y axis is an icon representing each student and the X axis is time. The software displays a mark next to the student icon at every time interval at which that students submits data. This allows for the identification of patterns in submissions. For example the student highlighted in the top box in Figure 6 has submitted 18 times during the activity period while the student in the lower box has only submitted 5 times.

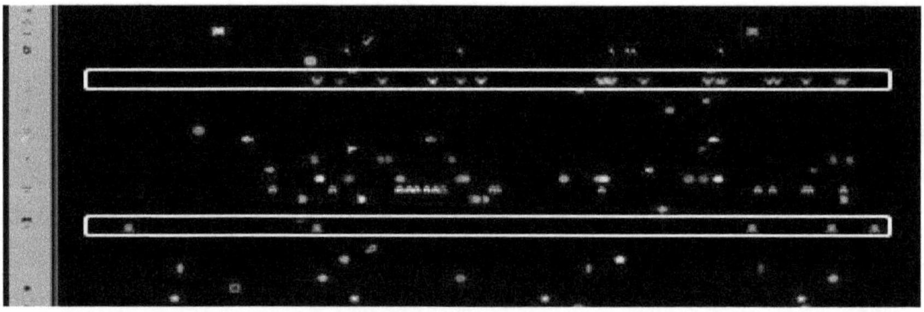

Figure 6. Timeline viewer class view

In addition to the whole class view, the Timeline Viewer lets you filter out other data and view the submissions from just one student (Figure 7). Using the placement of the icon as the timestamp (not the leading edge of the expression) the student view in Figure 7 shows that this student submitted a new Y1, Y2, Y3 and Y4 at the same time. Y1,

Y2 and Y3 were correct matches for the target expression in this activity which was Y=4X. When using the software correctness is indicated by the shading of the expression's text, as this paper is in black and white, boxes have been placed around the responses that match Y=4X. The software knows which match and which do not because of the identifying data added to each expression in the Graphical Viewer. The student then corrected Y4 and returned to Y1 to work at finding a new expression to replace 4X. In the end this student submitted 5 correct expressions.

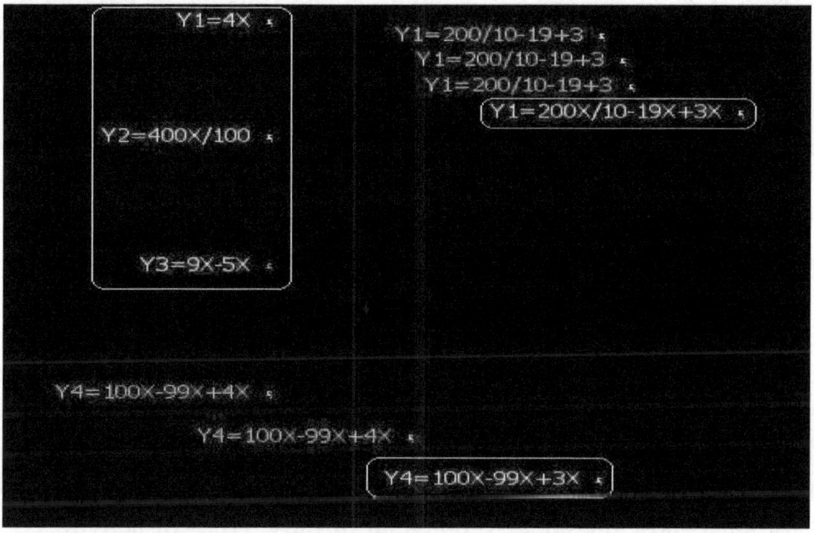

Figure 7. Timeline viewer individual student view

These two software interfaces have been a great first step to finding ways to visualize the large amounts of digital data but there is so much more that can be done to provide views of the data that teachers can use to make formative decisions about class and student knowledge. To get to a point where teachers can quickly see: What was the dominant mathematical strategy from class today? What mathematical strategies were used? Are there strategies that need to be reviewed? How is a particular student doing? Do they need help? Are they consistently struggling with a specific type of problem?

9 Conclusion

Problem solving can take many forms. This chapter has explored generative activities, a form of problematizing mathematics that utilizes the affordances of the full class to make meaning of the given tasks. The Generative Activities discussed in this chapter are rooted in the tradition of function-based algebra and utilize a classroom network of handheld devices. For the created curriculum, the algebraic topics covered in the Secondary 1 scheme of work were rearranged into three structural concepts; equals (where two expressions are everywhere the same), equivalence (the intersection of two expressions) and concepts of the linear functions (slope, rate, intercept). The activities were created using Generative Design principles. These were space creating play, dynamic structure, agency and participation. Generative Activities should open up a space for students to engage playfully with mathematics, to show creativity and interact with data sent in by other students. This playful space makes for a dynamic classroom experience where both teacher and students influence the course of the lesson. In giving students influence over the lesson content, their agency in the classroom is increased. Finally, the combination of the three previous principles gives students increased opportunity to participate in class. Results from the Singapore classrooms show the creative potential of these types of activities. The responses show both mathematical and social creativity. Many important mathematical concepts are generated by students and are available to the class to discuss. The teachers involved with the project have put in much hard work on changing their pedagogical practices to encourage creativity and take advantage of the artifacts in the group space to further students' conceptual understanding of mathematics. Generative Activities, coupled with concepts of function-based algebra and the affordances of a classroom network create a powerful problem solving environment.

References

Ball, D. L., & Bass, H. (2000). Interweaving content and pedagogy in teaching and learning to teach: Knowing and using mathematics. In J. Boaler (Ed.), *Multiple perspectives on the teaching and learning of mathematics* (pp. 83-104). Westport, CT: Ablex.

Boaler, J. (1998). Open and closed mathematics: Students experiences and understandings. *Journal for Research in Mathematics Education, 29*(1), 41-62.

Boaler, J. (2002). *Experiencing school mathematics: Traditional and reform approaches to teaching and their impact on student learning*. Mahwah, New Jersey: Lawrence Erlbaum Associates.

Davis, S. M. (2003). Observations in classrooms using a network of handheld devices. *Journal of Computer Assisted Learning, 19*(3), 298-307.

Davis, S. M. (2007). *Generative activities in Singapore: A beginning*. Paper presented at the International Conference of Teachers of Mathematical Modeling and Applications (ICTMA). Bloomington, IN.

Davis, S. M., & Brady, C. (2008a). GenSing Graphical Viewer.

Davis, S. M., & Brady, C. (2008b). GenSing Timeline Viewer.

Kaput, J. J. (1995). *A research base supporting long term algebra reform?* Paper presented at the Seventeenth Annual Meeting of the North American Chapter of the International Group for the Psychology of Mathematics Education, Columbus, OH., ERIC

Kaput, J. J. (2000). *Transforming algebra from an engine of inequity to an engine of mathematical power by "algebrafying" the K-12 curriculum*. Dartmouth, MA: National Center for Improving Student Learning and Achievement in Mathematics and Science. (ERIC Document Reproduction Service No. ED441664)

Lampert, M. (1990). When the problem is not the question and the solution is not the answer: Mathematical knowing and teaching. *American Educational Research Journal, 27*(1), 29-63.

Lampert, M. (2001). *Teaching problems and the problems of teaching*. New Haven, CT: Yale University Press.

Lesh, R. R., & Doerr, H. M. (2003). *Beyond constructivism: Models and modeling perspectives on mathematics problem solving, learning, and teaching*. Mahwah, NJ: Lawrence Erlbaum Associates.

Lloyd, G. M., & Wilson, M. S. (1998). Supporting innovation: The impact of a teacher's conceptions of functions on his implementation of a reform curriculum. *Journal for Research in Mathematics Education, 29*(3), 248-274.

National Council of Teachers of Mathematics. (2000). *Principles and Standards for School Mathematics*. Reston, VA: Author.

Stein, M. K., Baxter, J. A., & Leinhardt, G. (1990). Subject-matter knowledge and elementary instruction: A case from functions and graphing. *American Educational Research Journal, 27*(4), 639-663.

Stroup, W. M., Ares, N., Hurford, A. C., & Lesh, R. (2007). Diversity by design: The what, why and how of generativity in next-generation classroom networks. In R.A. Lesh, E. Hamilton, & J.J. Kaput (Eds.) *Foundations for the future in mathematics education* (pp. 367-394). Mahwah, NJ: Erlbaum

Stroup, W. M., Ares, N. M., & Hurford, A. C. (2005). A dialectic analysis of generativity: Issues of network supported design in mathematics and science. *Mathematical Thinking and Learning, An International Journal, 7*(3), 181-206.

Stroup, W. M., Carmona, L., & Davis, S. M. (2005). Improving on expectations: Preliminary results from using network-supported function-based algebra. In S. Wilson (Ed.), *Proceedings of the 27th annual meeting of the north American chapter of the international group for the psychology of mathematics education* [CD-ROM]. Blacksburg, VA.

Stroup, W. M., & Davis, S. M. (2005). Generative activities and function-based algebra. In H. L. Chick & J. L. Vincent (Eds.), *Proceedings of the twenty-ninth conference of the international group for the psychology of mathematics education* (Vol. 1, p. 328). Melbourne: PME.

Stroup, W. M., Kaput, J. J., Ares, N., Wilensky, U., Hegedus, S., Roschelle, J., et al. (2002). The nature and future of classroom connectivity: The dialectics of mathematics in the social space. In D. S. Mewborn, P. Sztajn, D. Y. White, H. G. Wiegel, R. L. Bryant & K. Nooney (Eds.), *Proceedings of the twenty-fourth annual meeting of the north American chapter of the international group for the psychology of mathematics education* (Vol. 1, pp. 195-213). Columbus, OH: ERIC.

Swafford, J. O., Jones, G. A., & Thronton, C. A. (1997). Increased knowledge in geometry and instructional practice. *Journal for Research in Mathematics Education, 28*(4), 467-483.

Vygotsky, L. S. (1978). *Mind in society: The development of higher psychological processes.* Cambridge, MA: Harvard University Press.

Wilensky, U. (1999). NetLogo. *The center for connected learning and computer-based modeling*. Northwestern University, Evanston, IL.

Chapter 9

Mathematical Modelling and Real Life Problem Solving

ANG Keng Cheng

Mathematical modelling is commonly regarded as the art of applying mathematics to a real world problem with a view to better understand the problem. As such, mathematical modelling is obviously related to problem solving. However, they may not mean the same thing. In this chapter, various aspects of mathematical modelling and problem solving will be discussed. Using concrete examples, some of the basic ideas and processes of mathematical modelling will be introduced and described as an approach to problem solving. In all the examples, a computing tool is used in part of the modelling process, demonstrating the critical role of technology in mathematical modelling. Some possible extensions of the modelling problems are also presented.

1 Introduction

Mathematical modelling has featured prominently in school mathematics curricula as well as many tertiary mathematics courses. However, despite this fact, mathematicians and mathematics educators alike, have not been able to reach a consensus on a precise definition of the term. It appears that different researchers adopt different definitions, depending on their field of work (Blum, 1993).

In fact, over the years, several interpretations of mathematical modelling arising from different perspectives and research directions have been proposed and used. For instance, as far as Cross and Moscardini (1985) and Bassanezi (1994) are concerned, mathematical modelling is defined simply as a process of understanding, simplifying and solving a real life problem in mathematical terms. However, according to Mason and Davies (1991), mathematical modelling is the movement of a physical situation to a mathematical representation.

Swetz and Hartzler (1991) define mathematical modelling as a process of observing a phenomenon, conjecturing relationships, applying and solving suitable equations, and interpreting the results. This seems to make mathematical modelling a rather scientific endeavour. In contrast, Yanagimoto (2005) thinks that mathematical modelling is not just a process of solving a real life problem using mathematics; it has to involve applying mathematics in situations where the results are "useful in society".

There are some researchers who hold the view that all applications of mathematics are mathematical models (Burghes, 1980). However, there are also those who feel that there is a difference between mathematical modelling and applications of mathematics (Galbraith, 1999). In fact, Galbraith claims that in a typical mathematical application, although the mathematics and the context are related, they are separable. In other words, after applying the necessary mathematics to solve the problem in some given context, we no longer "need" the context. A modelling task is distinctly different in that the focus is on investigating a particular problem or phenomenon, and the mathematics used is simply a means in understanding or solving the problem.

Whatever the views and differences in definition, one thing is clear: mathematical modelling has to have some connection with real life problems. Mathematical modelling is more than just problem solving; the problem to be solved arises from a real life situation, or a real life phenomenon. At times, the actual problem is not solved, but through the process of modelling, a better understanding of the problem is achieved.

1.1 *What is mathematical modelling*

Although there is no consensus on the precise definition of mathematical modelling, a working definition will be adopted for the purpose of this chapter. It is as follows. Mathematical modelling can be thought of as a process in which there is a sequence of tasks carried out with a view to obtaining a reasonable mathematical representation of a real world problem. Very often, in practice, this process is more like a cycle in which a model is continuously constructed, validated and refined. This process is illustrated in Figure 1 (Ang, 2006).

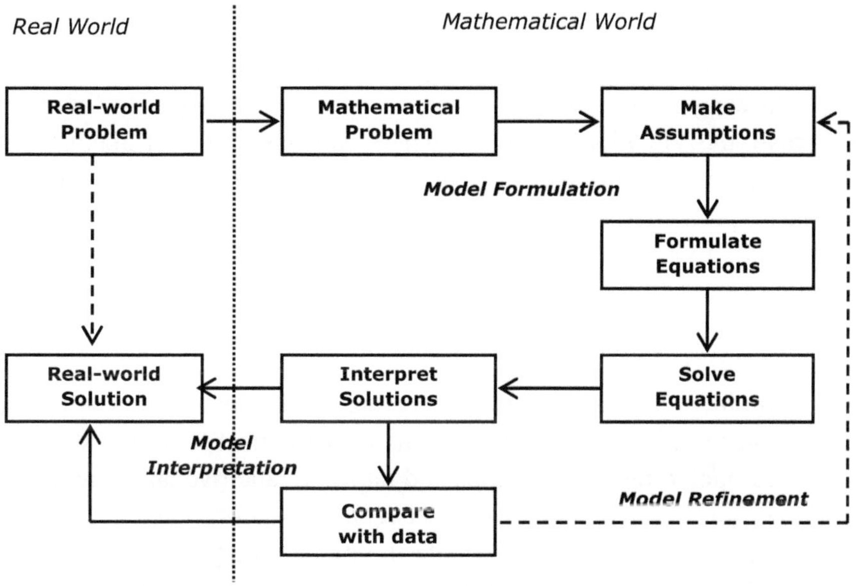

Figure 1. The mathematical modelling process

Beginning with a real life problem, the objective is to produce a real life solution. A direct approach to do this may be difficult or impossible. Thus, the first step in the mathematical modelling process is to understand the problem, and describe it in mathematical terms. In other words, *mathematize* the problem. In doing so, it is essential to be able to identify the variables in the problem, and to form relationships between or amongst these variables.

The next step is to construct a basic framework for the model. At this stage, some assumptions may need to be made. These assumptions are often necessary to keep the problem tractable, and simple enough to be solved by known methods. Of course, this means that the solution arising from the model can only be as good as its assumptions. Nevertheless, it is good to start with a simple model and then relax the assumptions later.

Given the assumptions, a model is constructed. The model may be an equation, a set of equations, a set of rules or simply an algorithm governing how values of the variables may be found or assigned. Generally, this is the most crucial stage and is often also the most difficult. It is also at this stage that the real physical meanings of the variables in the problem are used to justify the formulation of the model.

Following the formulation of the model, the next step is to find ways to solve the equations. This is where various different methods or strategies in problem solving can be exploited. In practice, unless a model is particularly simple, very often some kind of technological or computing tool will need to be used. The result from this step is a solution or a set of solutions to the mathematical problem that has been formulated.

The next step is to link the results or solutions of the model to the real world problem. This involves interpreting the results in physical terms. At this stage, it is common that various mathematical tools and skills are involved, including use of graphs and tables, qualitative and quantitative analyses, and so on. Comparisons between the solutions and collected or known data can be made to validate the model. Very often, a report on the results and interpretations becomes a "product" of this modelling process.

Although the modelling process may seem to terminate or culminate at the model interpretation stage, upon comparison with observed data, it may be possible to find ways to refine or improve the model. One common practice is to rethink the assumptions and perhaps modify or relax some of the earlier assumptions so that a more realistic or more reasonable model can be obtained.

From the above discussion, it is clear that mathematical modelling involves more than just a typical mathematical problem solving exercise.

In mathematical modelling, there is a clear and distinct connection to a real life problem.

1.2 *Problem solving examples from textbooks*

In order to appreciate the distinction between typical mathematical problem solving and typical mathematical modelling, it may be useful to examine how some local textbook writers perceive mathematical problem solving. Below are three typical problem solving exercises from textbooks or assessment books used by students in Singapore.

Problem Solving Item 1 (Upper Primary)

Paul had 30 more marbles than Peter. After Peter gave Paul 15 marbles, Paul had twice as many marbles as Peter. How many marbles did they have altogether?

Problem Solving Item 2 (Lower Secondary)

A cargo container is a cuboid that is 6.06 m long, 2.44 m wide and 2.59 m high.
(a) Find its total surface area (i) in m^2, (ii) in cm^2.
(b) Find its volume (i) in m^3, (ii) in cm^3.

Problem Solving Item 3 (Post Secondary)

A spherical balloon is being deflated in such a way that the volume is decreasing at a constant rate of 120 $cm^3\,s^{-1}$. At time t s, the radius of the balloon is r cm.
Find the rate of change of the radius when r = 30.
Find the rate of change of the surface area when the volume is $36\pi^4 cm^3$.

In Item 1, the expected method of solution is "model drawing method" (See Ministry of Education, 2009). It is clear that the actual context of the problem (that is, the marbles or the number of marbles that Paul and Peter own) is not really that important. What matters most is the method that can be used to tackle the problem, and, perhaps, the numerical computations that are required in the solution process.

Item 2 can be useful in testing the learner's ability to recall and use the formulae for the volume of a cuboid (and possibly also the meaning of the term) and the area of a rectangle, and the ability to perform unit conversion. Whether the cuboid is a cargo container or not is immaterial and not at all important. In other words, one could have just said "a cuboid is …" and the item and its solution would have remained exactly the same.

Item 3 expects the use of the concept of rates of change in calculus. Once it is recognized that the object is spherical (and the formula for the volume of a sphere is recalled), the fact that it is a balloon that is deflating is no longer of any importance.

In all the above typical problems, the focus is on applying or using some specific mathematical concept or skill to solve the problem. One could generate many such "problems" using different contexts without changing the intent of the exercise. In other words, these problem solving items focus on the use of mathematics, rather than the context or the "problem". These typical textbook problems, strictly speaking, could hardly be called "real life" problems although some authors do try to inject some real life element into the context.

Mathematical modelling, on the other hand, focuses on the problem itself. Typically, the problem concerns real systems or real problems that one could possibly cast into a mathematical context and attempt a solution. If the context or problem is changed, one would probably need to use a different solution technique or approach.

In the next section, we examine the different approaches to mathematical modelling.

2 Approaches to Mathematical Modelling

There are several different ways in which one can employ a mathematical model to solve real problems. These different ways may be classified into four broad approaches to mathematical modelling.

2.1 *Empirical models*

In empirical modelling, one examines data related to the problem. The main idea is to formulate or construct a mathematical relationship between the variables in the problem using the available data. Typically, one uses methods such as the method of least squares, or some other kind of approach that minimizes the error between the observed data and the modelled relationship.

In this approach, the model usually involves certain unknown parameters that need to be obtained or estimated from the data set. The main advantage is that the resulting model is capable of reproducing the data set in an accurate manner. This approach is also very simple and easy to apply in most cases. There are many technological tools that can do curve fitting efficiently.

However, one main disadvantage of empirical modelling is its over-reliance on historical data. One cannot be sure if the same model is still applicable outside the range of the data set used. In other words, while it can be used to explain historical relationship, it may not be useful for predictions.

Another shortcoming is that parameters in empirical models often are just numerical values which may not have any real physical meaning in the problem. Given a different data set, these parameter values will be different.

2.2 *Simulation models*

Simulation models involve the use of a computer program or some technological tool to generate a scenario based on a set of rules. These rules arise from an interpretation of how a certain process is supposed to evolve or progress.

Typically, simulation is used to model a phenomenon or situation when it is either impossible or impractical to conduct physical experiments to study it. For instance, one may simulate a certain design for a telecommunication network to find the best design. It would be too expensive to build an actual system to test the design. Using a simulation, one could test how the network performs at different traffic loads, or whether a particular routing algorithm could increase performance levels, and so on.

Simulation can be discrete events models or continuous models. In discrete events, the assumption is that the system changes instantaneously in response to changes in certain discrete variables. Continuous simulations, on the other hand, changes are continuously fed into the system over time and responses are continuously quantified.

2.3 *Deterministic models*

Generally, when we use an equation, or a set of equations (which may include ordinary differential equations, partial differential equations, integral equations, and so on), to model or predict the outcome of an event or the value of a quantity, we are using deterministic models. The equation or set of equations in a deterministic model represents the relationship amongst the various components or variables of a system or a problem.

For instance, the equations of motion, based on Newton's laws, are a set of deterministic models governing the motion of a particle. Thus, when a ball is tossed up in the air, given that certain variables (such as its initial velocity) are known, we can use the model to predict its motion at a later time. It is important to note that there are some assumptions that must be stated when using models. In this case, the ball is assumed to behave like a particle, and air resistance is assumed to be negligible, and so on.

2.4 *Stochastic models*

In deterministic modelling, random variations are ignored. In other words, the equations used to represent real world problems are formulated based on fundamental relationships between the component variables in the problem. Generally, one set of conditions results in one solution.

Many real world problems, however, are subjected to random variations and fluctuations. As an example, consider the modelling of chemical reactions. Although it is possible to construct equations to predict the behaviour of reacting substances, chemical reactions occur only if there are effective molecular collisions. In other words, there is some degree of randomness and uncertainty. Thus, different outcomes may arise from the same set of initial conditions.

In stochastic models, randomness and probabilities of events happening are taken into account when the equations are formulated. The model is constructed based on the fact that events take place with some probability rather than with certainty. In recent years, such stochastic models have become very popular with researchers and professionals in the fields of finance, business and economics.

3 Examples

In this section, we illustrate the different approaches of mathematical modelling through examples in real life applications. It should be noted that practical mathematical modelling often requires the use of some technological tools. Describing these tools in detail is beyond the scope of the current discussion although some simple technique with regard to using the *Solver* function of the spreadsheet, MS Excel, will be mentioned. The reader may wish to refer to the relevant references such as Beare (1996) for details.

Example 1: Modelling elastic blood vessels

In the mathematical modelling of blood flow through elastic arteries, it is necessary to obtain a relationship between the stress (tension, $T(x)$) experienced by an elastic material caused by the strain (x) exerted on it (see Mazumdar, Ang and Soh, 1991). There are different ways of modelling this relationship. In this example, we make use of empirical data obtained experimentally by Roach and Burton (1957) for an iliac artery. The set of data is reproduced in Table 1 below and plotted in Figure 2.

Table 1

Observed experimental values of stress and strain in an iliac artery

strain (x)	stress (y)
1.22	4
1.35	8
1.45	13
1.50	20
1.55	22
1.57	28
1.60	33
1.64	40
1.67	44
1.71	60
1.74	71
1.77	83
1.80	95
1.83	109

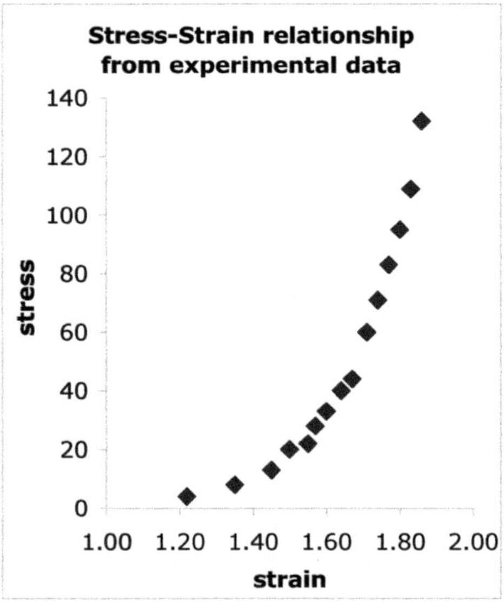

Figure 2. Graph of stress against strain in iliac artery

The relationship between the stress and strain in an elastic artery is assumed to take the form

$$T(x) = Ae^{kx} + B\,, \tag{1.1}$$

where A, B and k are constants to be determined. Applying the condition for an unstressed blood vessel, namely, $T\ (1) = 0$, the equation can be simplified to

$$T(x) = A(e^{kx} - e^{k})\,, \tag{1.2}$$

where $x \geq 1$, and A and k are both positive. The values of parameters A and k may be estimated using the least squares method, or by minimizing the error between the model and the data. Defining the "sum of residual squares" (SRS) as

$$S = \sum_{i=1}^{n} (y_i - T(x_i))^2\,, \tag{1.3}$$

where (x_i, y_i) are the observed data points, a spreadsheet may be used to find the values of A and k that will minimize S. In Microsoft Excel, for instance, the "solver" tool may be used. The Solver tool allows the user to minimise (or maximise) the value of a selected cell by varying the values of other cells specified by the user. In the present case, the Solver tool returns the value of $k = 0.007427$ and $A = 5.263$ (to four significant figures) with a minimum value of $S = 39.53936$. These results from the spreadsheet are shown in Table 2, along with a graph of the model and experimental values in Figure 3.

Empirical modelling can be easily exploited by the mathematics teacher in the classroom. What is required would be an interesting data set, some knowledge of functions and their graphs and a good technological tool capable of performing function approximations.

Table 2

Results of fitting model to data using Solver tool in Microsoft Excel

	***A* =**	**0.007427**	
	***k* =**	**5.263085**	
Strain	Stress		Squared Error
x	y	$T(x)$	
1.22	4	3.13053	0.75598
1.35	8	7.61369	0.14924
1.45	13	13.88072	0.77567
1.50	20	18.49085	2.27752
1.55	22	24.48876	6.19393
1.57	28	27.36621	0.40168
1.60	33	32.29220	0.50098
1.64	40	40.19504	0.03804
1.67	44	47.31528	10.99109
1.71	60	58.73839	1.59166
1.74	71	69.03029	3.87977
1.77	83	81.08251	3.67675
1.80	95	95.19616	0.03848
1.83	109	111.72381	7.41912
1.86	132	131.07835	0.84944
		***S* =**	**39.53936**

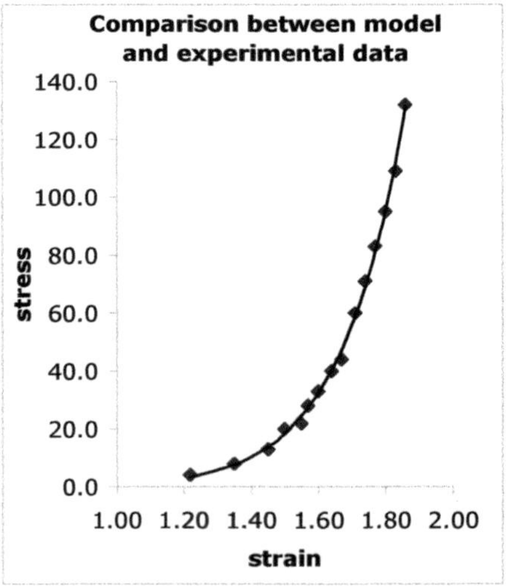

Figure 3. Graph of model and experimental data

For instance, students could be asked to collect data on the growth of bean sprouts over, say, a three-week period and then construct a mathematical model for the growth based on their data. As another example, the teacher could ask students to find information about "braking distances" of vehicles from the Internet. These are usually experiments carried out by automobile companies to test the braking efficiencies of their vehicles. The challenge is to construct a model that will predict the braking distances of certain makes of cars.

Example 2: Random Walk

For an example of the simulation model, consider a "random walk" problem. Suppose a person begins walking at some starting point. Being quite drunk, she takes a step in north, south, east and west directions with equal probability. That is, there is an equal chance that she takes a step in any of the four directions. The problem is to determine how far from the starting would she have gone after taking, say, 100 steps.

One possible way to look at this problem is to use some computing tool to simulate the situation. In this case, each step taken by the person is regarded as an event, and we make the assumption that the events are independent. This means that the person's next step will not be dependent on her previous step. Since each event is a separate distinct event, the simulation model is a discrete event simulation.

Again, a spreadsheet such as Microsoft Excel may be used to construct this simulation model. The spreadsheet can be set up with three columns. The first two columns contain the x and the y coordinates of the position of the person respectively, and the third column stores the direction in which she will move in the next step. The directions can be conveniently coded as "1", "2", "3" and "4" to represent "East", "North", "West" and "South" respectively. Figure 4(a) below shows the first 10 rows of a truncated Excel worksheet set up for the random walk simulation.

The values in cells A2 and B2 are set to 0, indicating that the walk begins at the origin (0, 0). In cell C2, the direction for the next step is simulated by using the Excel formula "= randbetween(1,4)", which returns an integer between 1 and 4, and randomly drawn from a uniform distribution. Thus, each of the integers 1, 2, 3 and 4 gets an equal chance of being selected. Depending on the value assigned to cell C2, the values in cells A3 and B3 will be assigned accordingly. The rules of the random walk are translated to the following algorithm:

> If value in C2 = 1, then: A3 = A2 + 1. If value in C2 = 3, then: A3 = A2 – 1.
>
> If value in C2 = 2, then: B3 = B2 + 1. If value in C2 = 4, then: B3 = B2 – 1.

In Excel, these may be implemented by entering the following formulae in cells A3 and B3 respectively:

"`=IF(C2=1,A2+1,(IF(C2=3,A2-1,A2)))`" and

"`=IF(C2=2,B2+1,(IF(C2=4,B2-1,B2)))`".

The formulae in cells C2, and cells A3 and B3 are then copied to cells directly below until 100 steps are taken. A typical run of this simulation is shown graphically in Figure 4 (b).

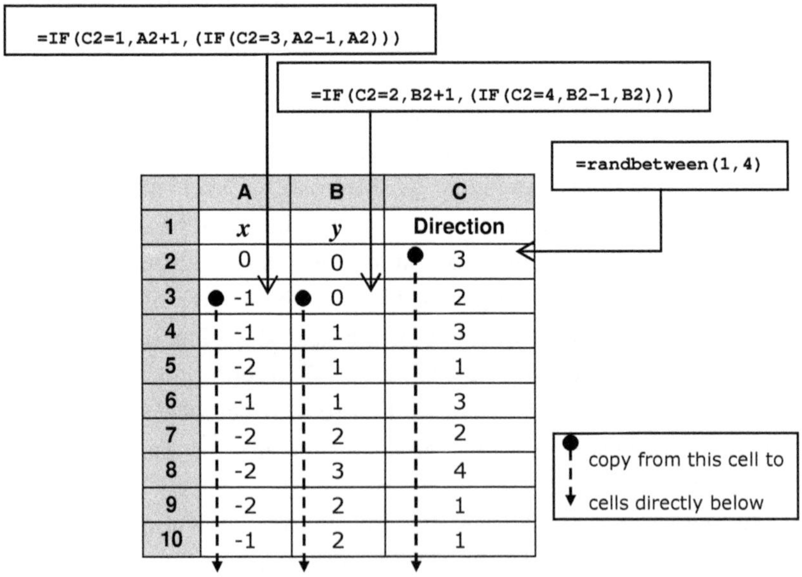

Figure 4(a). First 10 rows of spreadsheet in Random Walk example

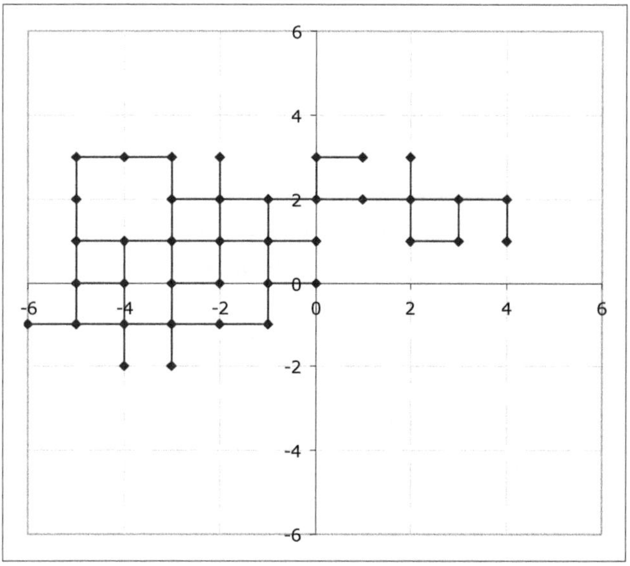

Figure 4(b). A typical graphical output of random walk simulation

Some possible extensions include changing the probabilities of the choosing each direction for the next step in the random walk. Another possibility is to impose other rules such as taking 3 steps in one direction before changing direction.

Although it may seem rather pointless to simulate such a situation, in actual fact, this simulation model can be applied to model movement of cells in a tissue. Another application is in the modelling of growth of cells, or abnormal growth of cells leading to tumour formation.

As a classroom activity, one suggestion is to start with a one-dimensional random walk. That is, the walk is restricted to just moving along the, say, x-axis. At any one point, the probability of moving left at the next step is equal to that of moving right. This one-dimensional random walk can be simulated and it will not be difficult for students to with access to a tool like Microsoft Excel to construct the simulation model. This activity leads very naturally to binomial distribution, and then, as the number of steps grows, to the normal distribution.

Example 3: A disease outbreak (Logistic model)

A basic model for the spread of an infectious disease can be constructed from a simple first order differential equation. In this model, healthy and susceptible individuals who come into contact with infected and infectious individuals will themselves get infected. In other words, there exists a movement of members from one compartment ("Susceptible") to another compartment ("Infected"). For this reason, such a model is often called the "S-I" infectious disease model.

Assuming a closed community with a total of N individuals, and denoting the number of infected individuals by x, the S-I model can be written as the differential equation,

$$\frac{dx}{dt} = kx\left(1 - \frac{x}{N}\right) \tag{1.4}$$

where k is some constant related to the rate of transmission of the disease. This equation is also commonly known as the logistic equation. The solution to the logistic equation is

$$x = \frac{N}{1 + (N / x_0 - 1)\, e^{-kt}}, \tag{1.5}$$

where $x_0 = x(0)$ is the number of infected individuals at the start of the outbreak. Although this is a very simplistic view of a disease outbreak, it can be applied to a real situation.

In the year 2003, a deadly disease struck some parts of the world. The disease, now known as Severe Acute Respiratory Syndrome, or SARS, was an emerging infectious disease that spread very rapidly. Thousands of cases were reported, and hundreds had died. SARS had struck about 30 countries and in Singapore, 206 cases were recorded and 31 infected people lost their lives during the 70-day outbreak.

Data for the SARS outbreak in Singapore are available in the public domain (Heng and Lim, 2003) and are reproduced in Table 3. Suppose $x(t)$ represents the number of infected individuals at time t, measured in days. From the data, it is clear that $x_0 = x(0) = 1$. That is, the outbreak had started with just one infected individual. Also, assuming that the community is closed (which means that no one enters or leaves the system during the outbreak period), the total number of individuals is assumed to be $N = 206$. This is, of course, a debatable assumption. However, without this assumption, it would not be possible to use this model. Moreover, the idea here is to test and see if the logistic equation serves well as a reasonable deterministic model for the SARS outbreak.

Table 3

Number of individuals infected with SARS during the 2003 outbreak in Singapore (Heng and Lim, 2003)

Day (t)	Number (x)	Day (t)	Number (x)	Day (t)	Number (x)
0	1	24	84	48	184
1	2	25	89	49	187
2	2	26	90	50	188
3	2	27	92	51	193
4	3	28	97	52	193
5	3	29	101	53	193
6	3	30	103	54	195
7	3	31	105	55	197
8	5	32	105	56	199
9	6	33	110	57	202
10	7	34	111	58	203
11	10	35	116	59	204
12	13	36	118	60	204
13	19	37	124	61	204
14	23	38	130	62	205
15	25	39	138	63	205
16	26	40	150	64	205
17	26	41	153	65	205
18	32	42	157	66	205
19	44	43	163	67	205
20	59	44	168	68	205
21	69	45	170	69	205
22	74	46	175	70	206
23	82	47	179		

With these values, what is needed to complete the model is a value for the transmission rate, k. One way to find an estimate for k is to use the available data, and find a curve of best fit. To do so, an "average error", E, defined as

$$E = \frac{\sqrt{\sum_{i=1}^{n} (\hat{x}_i - x_i)^2}}{n}, \qquad (1.6)$$

where $\hat{x}_i$ are the data values, x_i are the values obtained from the model, and $n = 71$ is the total number of data points available, is used. As before, if a spreadsheet like Microsoft Excel is used, then the "Solver" tool may be used to find the value of k that minimizes the error E. For this data set, it turns out that the minimum value of E is found to be 1.9145 when $k = 0.1686$. Results of this modelling exercise is shown in Figure 5 below.

From the graphs in Figure 5(b), it can be seen that the model does not give a very good fit to the data. It appears that at the beginning and at the end of the outbreak, the model appears to be fairly reasonable. However, between t=15 and t=50, the model deviates from the actual SARS cases quite significantly.

In fact, it is possible to refine the model so that a better fit can be obtained. The logistic equation assumes a linear relationship between the fractional rate of change of $x(t)$ with $(1 - (x/N))$. A more general logistic model would be to relax this assumption so that the fractional rate of change of $x(t)$ varies with $(1 - (x/N)^p)$ for some real constant p. The same procedure for finding the new value of k and the value of p that will minimize the error E can be applied. The result is a modified or generalized logistic *model* for the SARS outbreak and is shown in Figure 6.

It is clear from Figure 6 that this new model is an improvement over the previous. In fact, the model can be further refined and improved. For details on how this can be done, the reader may refer to Ang (2004).

k =	**0.1686**	**E =**	**1.9145**
Day	**SARS cases**	**Logistic Model**	**Squared Error**
0	1	1.0000	0.0000
1	2	1.1826	0.6682
2	2	1.3982	0.3621
3	2	1.6529	0.1205
4	3	1.9536	1.0950
5	3	2.3083	0.4785
⋮	⋮	⋮	⋮
⋮	⋮	⋮	⋮
⋮	⋮	⋮	⋮
70	206	205.6838	0.1000

(a) Minimizing error using a spreadsheet

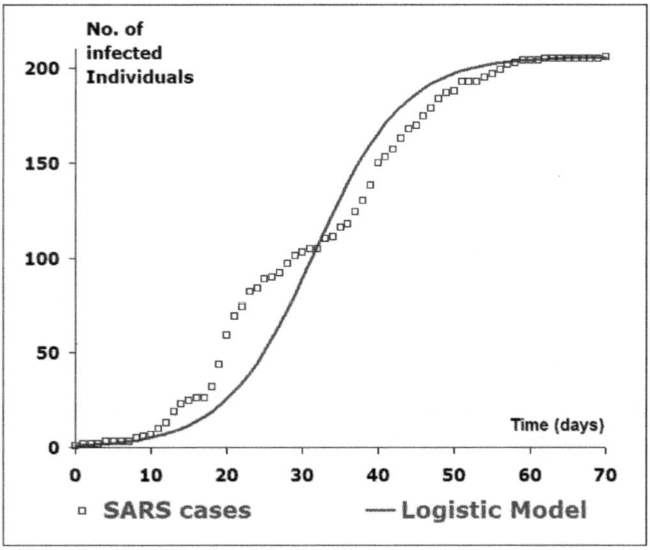

(b) Graph of SARS cases and solution from model

Figure 5. Modelling the SARS outbreak in Singapore using the logistic equation

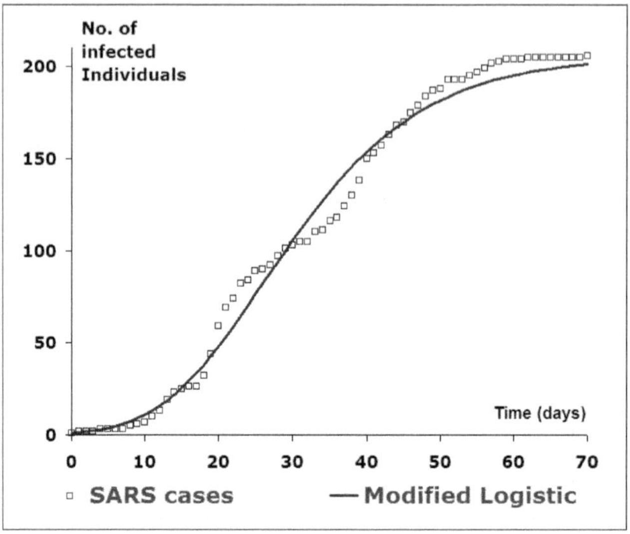

Figure 6. Modified logistic model with $p = 0.1988$ and $k = 0.4334$

This example illustrates the idea of model refinement in mathematical modelling. It also shows that even in deterministic modelling, one often needs to use empirical data to estimate parameters, such as k in this case. The difference is that in this case, the parameter k has a physical meaning, unlike in typical empirical models where parameters may sometimes not have any real physical meaning in the model.

Mathematical modelling applied to a local context tends to add authenticity to the task and arouse greater interest amongst students. This is the reason why Example 3 has received much attention from local teachers when it was first discussed. Teachers may wish to look for relevant and real resources when sourcing for ideas.

Example 4: A disease outbreak (Stochastic model)

The same logistic model discussed in the preceding example may be further modified to include a stochastic term, leading to a stochastic or, perhaps more accurately, hybrid model for the SARS outbreak.

Rewriting the logistic equation with a stochastic term results in a stochastic differential equation (SDE) given by

$$dX(t) = \lambda\, X(t)\,(N - X(t))\,dt + \mu\, g(X(t))\,dW(t)\,,$$

$$X(0) = X_0\,, \quad 0 \le t \le T\,. \tag{1.7}$$

In this equation, λ is a constant, $W(t)$ is a random variable representing a standard Wiener process and μ is a scaling factor. Here, the deterministic portion of the equation, namely $\lambda\, X(t)\,(N - X(t))$ is commonly known as the "drift" and the stochastic part $\mu\, g(X(t))\,dW(t)$ is known as the "diffusion" term. The constant, λ, as before, is related to the transmission rate of the disease. The function $g(X(t))$ is used to govern the dependence of the $X(t)$, that is the number of infected individuals, on the "noise" or uncertainties. For instance, to model a simple linear relationship between $X(t)$ and the stochastic term, one could set $g(X) = X$.

The above equation can be solved using the Euler-Mayurama numerical method. In this method, the equation is written in the discretized form

$$X_j = X_{j-1} + \lambda\, X_{j-1}(N - X_{j-1})\Delta t + \mu\, g(X_{j-1})\,(W_j - W_{j-1})\,,$$

$$\text{for } j = 1,\ 2, \ldots \tag{1.8}$$

To solve this equation using this method, a Brownian path needs to be generated so that the difference $(W_j - W_{j-1})$ can be given a value. In numerical simulations, it is usual to consider a discretized Brownian motion, in which $W(t)$ is sampled at discrete t values. Details of the derivation and solution method can be found in Ang (2007), in which the method is implemented through a program written in MATLAB.

The derivation of the stochastic model and the computer programming involved are beyond the scope of the current discussion. Interested readers may wish to refer to Ang (2007) for details on the actual construction and solution of this model. Video clips demonstrating the working of the MATLAB programs are found in http://math.nie.edu.sg/kcang/ejmt0702.

4 Pedagogical Implications

Apart from solving routine mathematical problems in a context-free environment, it is useful to consider real life applications of mathematics. Typical textbook problems on "real life applications" present problems in a very clean and tidy state. Such practice makes it difficult to convince the learner that real life applications of mathematics do indeed exist. Mathematical modelling provides an avenue for teachers and their learners to look at problem solving from a problem-centred point of view.

It is not uncommon to find students thinking of mathematics as consisting of a set of distinct topics that are compartmentalized and self-sufficient. Real life problems tend to transcend a number of disciplines and are often not so well defined. Often, one needs to apply ideas and concepts in one area to solve problems arising in another. Mathematical modelling offers excellent opportunities to connect and use ideas from different areas.

To teach and learn mathematical modelling successfully, some skills and understanding of the processes involved in the model are required. It is not easy for a teacher to have the same kind of experience or skills that a professional applied mathematician would have acquired over time. However, it is possible for teachers to learn alongside their students.

It is important to note that there is a difference between teaching mathematical modelling and mathematical models. In the latter, the emphasis is on the product (the models). In mathematical modelling, the focus is on the process of arriving at a suitable representation of the physical, real world problem and solution. It is important that the teacher is mindful of this difference and be prepared to accept a situation where no solution or multiple solutions are reached.

As can be gleaned from the examples discussed in the preceding section, mathematical modelling provides an excellent platform for studies and experiments of an inter-disciplinary nature. Problems often arise in a different discipline and this provides the mathematics teacher with excellent opportunities to collaborate with other teachers in problem-solving.

It should also be noted, through the examples presented, that the use of technology plays a critical role in mathematical modelling. In real life problems, one often has to deal with real life data, which may not be as clean or "sanitized" as textbook examples or data. In such instances, rather than struggle with complicated or tedious numerical computations, it may be better to use a tool so that one could focus on the mathematics.

In some cases, technology can also help make the mathematics more accessible. For instance, in our example, we have used a feature in a spreadsheet tool that helps us to find parameters that best fits a function to a set of data. It is possible to work out the parameters by hand, using mathematics that may be a little too advanced for the learners for which the problem was originally intended. The use of this technological tool thus bridges the gap, which the student can fill in good time.

5 Concluding Remarks

In this chapter, the use of mathematical modelling as a means of problem solving is examined. While problem solving items in textbooks can provide the learner with ample opportunities to hone the necessary mathematical skills in problem solving, real problems provide a rich context in which the learner can actually use or apply these skills in a real context. The experience will be even more enriching if the problem involves issues of public concern (such as spread of a disease like dengue, modelling traffic flow in a city with traffic problems, and so on).

It is also important to recognize and acknowledge that while one may attempt to solve the problem, in practice, it will not be surprising if one fails to completely solve the problem. Moreover, mathematical models can only be as good as their assumptions. Real life problems may need to be simplified to make them more tractable and manageable. Nevertheless, tackling a complex real life problem through mathematical modelling will be an enriching problem solving experience for both teacher and learner.

References

Ang, K.C. (2004). A simple model for a SARS epidemic. *Teaching Mathematics and Its Applications*, 23(4), 181–188.

Ang, K.C. (2006). *Differential equations: Models and methods*. Singapore: McGraw-Hill.

Ang, K.C. (2007). A simple stochastic model for an epidemic — numerical experiments with MATLAB. *The Electronic Journal of Mathematics and Technology*, 1(2), 116–127.

Bassanezi, R.C. (1994). Modelling as a teaching-learning strategy. *For the Learning of Mathematics*, 14(2), 31–35.

Beare, R. (1996). Mathematical modelling using a new spreadsheet-based system. *Teaching Mathematics and Its Applications*, 15(1), 120–128.

Blum, W. (1993). Mathematical modelling in mathematics education and instruction. In T. Breiteig, I. Huntley & G. Daiser-Messmer (Eds.), *Teaching and learning mathematics in context* (pp. 3–14). London: Ellis Horwood.

Burghes, D. (1980). Mathematical modelling: A positive direction for the teaching of applications of mathematics at school. *Educational Studies in Mathematics*, 11, 113–131.

Cross, M. & Moscardini, A.O. (1985). *Learning the art of mathematical modelling*. Chichester: Horwood and Wiley.

Galbraith, P. (1999). *Important issues in applications and modelling*. Paper presented at the AAMT Virtual Conference 1999, Adelaide, Australia.

Heng, B.H. & Lim, S.W. (2003). Epidemiology and control of SARS in Singapore. *Epidemiological News Bulletin*, 29, 42–47.

Mason, J. & Davis, D. (1991). *Modelling with mathematics in primary and secondary schools*. Geelong, Australia: Deakin University Press.

Mazumdar, J.N., Ang, K.C. & Soh, L.L. (1991). A mathematical study of non-Newtonian blood flow through elastic arteries. *Australasian Physical and Engineering Sciences in Medicine*, 14(2), 65–73.

Ministry of Education. (2009). *The Singapore Model Method for learning mathematics*. Singapore: Author.

Roach, M.R. & Burton, A.C. (1957). The reason for the shape of the distensibility curves of arteries. *Canadian Journal of Biochemistry and Physiology*, 35, 681–690.

Swetz, F. & Hartzler, J.S. (1991). *Mathematical modelling in the secondary school curriculum*. Reston, VA: The National Council of Teachers of Mathematics.

Yanagimoto, T. (2005). Teaching modelling as an alternative approach to school mathematics. *Teaching Mathematics and Its Applications*, 24(1), 1–13.

Part III

Mathematical Problems and Tasks

Chapter 10

Using Innovation Techniques to Generate 'New' Problems[1]

Catherine P. VISTRO-YU

Teachers sometimes run out of ideas and have difficulty creating new problems for our students to solve. When those times come, the remedy is to innovate on existing and old problems that have proven to be useful and effective in teaching mathematical skills. In this paper, the author gives some pointers on how to innovate on "used" problems with the objective of developing "new" problems to give to students. The technique is a borrowed concept from literature and is applied to mathematical problem solving, particularly, problem generation or problem formulation.

1 Introduction

Problem solving plays a very important role in the learning of mathematics. For one, problem solving develops higher-order thinking skills that we need to function in today's world (NCTM, 2000). Over the years, problem solving has truly been considered in three ways (Branca, 1980): as a goal (Singapore Ministry of Education, 2006a; 2006b), as a process (Polya, 1945), and as a basic skill (Malone, Douglas, Kissane and Mortlock, 1980; Schoen and Oehmke, 1980). Consequently, its inclusion in the curriculum or its teaching has taken on various forms. For school mathematics teachers, problem solving is, and must be, a

[1] Part of this paper was given in a workshop held at the Mathematics Teachers Conference 2008, National Institute of Education, Singapore.

staple ingredient in their mathematics lessons. However, problem solving cannot commence unless there is a problem to solve and a good problem to solve at that! Where does one get good problems?

One aspect of problem solving that school mathematics teachers need to engage in is the "art of problem posing" (Brown and Walter, 1983). Brown and Walter (1983) argued that one's level of mathematical understanding is closely linked to one's ability to generate and pose questions. Various studies have given numerous ideas and suggestions on how school mathematics teachers could develop their problem posing skills (e.g. Crespo, 2003; Yeap and Kaur, 1999; and Silver, Mamona-Downs, Leung, and Kenney, 1996) by either providing a framework for teachers to work with or identifying specific steps that teachers could follow in formulating both old and new problems. Indeed, problem generation or problem formulation is am important skill that mathematics teachers need to develop. In particular, Crespo and Sinclair (2008) consider problem posing as necessary for prospective teachers because teaching entails posing good questions that would aim for students' development of mathematical understanding. Problem posing serves as an excellent way for teachers to practice posing good questions, a necessary ingredient for generating excellent problems.

The technique that is described in this paper is what I call 'innovation', a concept borrowed from literature. The intent is to generate problems out of existing problems. Silver and his colleagues (1996) call it 'problem reformulation' while Crespo (2003) calls it 'adaptation' (p. 250). All three terms refer to the same technique of generating new problems out of old, existing problems except for some subtle differences.

2 Innovation: A Borrowed Idea from Literature

Corbett (2007) describes 'innovating on a story' as a technique (*http://www.teachit.co.uk/custom_content/newsletters/newsletter_jan07.asp#1*) in literature to develop story-telling and story-writing skills among children. Prior to innovation, a teacher chooses a story and tells it to the children. The teacher then engages the children in activities to help them internalize the story, its characters, setting, patterns of events, and then asks the children to tell the story themselves. Once the story has been internalized, innovation can be done. Corbett (2007) describes the innovation in the following manner:

> Once the children know the story really well and it is in their long-term working memory, then you can move on to 'innovation'. Initially, the children adapt their story map or board, making decisions – and then they try telling their new story. They will need to retell a number of times, refining their expression until they have orally redrafted to their satisfaction.

There are different types of techniques in innovating on stories and can range from very simple to complex types:

- substitution – retelling the same story but making a few simple changes such as names, objects, places
- addition – retelling the same story but adding in more description, dialogue or events
- alteration – making changes that have repercussions, e.g. altering characterization, modernising the setting, changing the ending
- change of viewpoint – retelling the story from a different character's view
- transformation – retelling the story in a different genre
- recycling the plot – re-using only the underlying plot pattern.

'New' stories created from innovating on existing stories show varying levels of creativity of students. Some stories do turn out to be mere copycats of the original story but others provide excitement with some new additions or twists in the plot making story innovation a useful tool in literature.

3 Innovation on Problems in Mathematics

The aim of problem posing by teachers is to generate good and mathematically valuable problems, not just any problem. Research have shown that problems generated based on a certain prompt or "outside of the context of inquiry" (Crespo and Sinclair, 2008, p. 397) are not always good or mathematically interesting problems. In fact, the study by Crespo and Sinclair (2008) showed that prospective teachers, who generated problems by first exploring the mathematical situation out of which they were to generate problems, were more successful in posing reasoning problems compared to the prospective teachers who posed problems spontaneously. The latter group produced more factual problems that are not mathematically interesting. Exploration is analogous to the prerequisite for a successful innovation in story-telling. Recall that students have to be able to internalize the original story before they could be expected to innovate. Crespo and Sinclair (2008) seem to be onto the same idea in problem posing. By exploring a mathematical situation, problem posers are able to understand better and internalize the situation. This certainly contributes to one's ability to pose problems that are more meaningful and mathematically valuable and interesting.

Crespo (2003) identified three approaches that preservice mathematics teachers use to pose problems to pupils in a study that used letter-writing as the mode of communication between preservice teachers and students. The three approaches are:

- making problems easy to solve;
- posing familiar problems;
- posing problems blindly.

It is in the first approach that preservice teachers tended to use adaptation as a way to generate problems.

Silver et al (1996) noted that some middle school mathematics teachers and preservice secondary school mathematics teachers generated problems by:

- keeping the problem constraints fixed and focusing their attention on generating goals ('accepts the given' by Brown & Walter, 1983);
- manipulating the given constraints of the task setting as they generated goals ('challenging the given' by Brown & Walter, 1983).

Table 1

Comparison of Innovation Techniques between Storytelling and Mathematic Problem Generation

Innovation on Stories	Innovation on Mathematics Problems	Feature of the Problem
substitution – retelling the same story but making a few simple changes such as names, objects, places	*replacement* – posing the same problem but changing quantities, amounts, units, shapes, etc.	Problem becomes a drill exercise.
addition – retelling the same story but adding in more description, dialogue or events	*addition* – posing the same problem but adding a new given or constraint or adding an obstacle	Problem is extended and could become more complex.
alteration – making changes that have repercussions, e.g. altering characterization, *modernising* the setting, changing the ending	*modification* – takes the same given but modifies the problem	Problem could become totally new but could still be solved using the original problem as a take-off point.
transformation – retelling the story in a different genre	*contextualizing* the problem to make it more relevant to students	Problem becomes more relevant but is basically the same problem as the original.
change of viewpoint – retelling the story from a different character's view	*turning the problem around* or *reversing the problem* – taking the same problem but taking the end goal as the given and the given as the end goal	Problem becomes more interesting and challenging and completely different.
recycling the plot – re-using only the underlying plot pattern	*reformulation* – posing the same problem in a different type (e.g. from a proving problem to a situational problem, see Butts, 1980)	Problem is different but uses knowledge of the same concept or skill as required from the original problem.

I believe there are more ways to generate problems by applying the innovation techniques in literature to problems in mathematics. Below is the corresponding set of innovation techniques as adapted from innovation techniques in storytelling:

- replacement – posing the same problem but changing quantities, amounts, units, shapes, etc.
- addition – posing the same problem but adding a new given or constraint or adding an obstacle
- modification – takes the same given but changes the problem
- contextualizing the problem to make it more relevant to students
- turning the problem around or reversing the problem – taking the same problem but taking the end goal as the given and the given as the end goal
- reformulation – posing the same problem in a different type (e.g. from a proving problem to a situational problem, see Butts, 1980)

Depending on the innovation technique used, the new problem generated may be better, worse, or just the same in terms of the level of difficulty, sophistication, and novelty. Table 1, shows the corresponding innovation techniques between storytelling and mathematics problem-generation and perceived features of problems resulting from the innovation technique.

In the next two sections, I discuss two problems and the new problems generated from them by applying the above-mentioned innovation techniques.

4 Using Innovation to Generate Problems

This section illustrates the use of innovation to generate problems through two examples.

4.1 *Example 1*

The Problem:

A merchant buys his goods at 25% off the list price. He then marks the goods so that he can give his customers a discount of 20% on the marked price but still make a profit of 25% on the selling price. What is the ratio of marked price to list price? (Krulik and Rudnick, 1989, p. 149).

A short solution to this problem is as follows:

> Let L be the list price. The merchant bought the goods at the price of $0.75L$.
> Let M be the marked price. The merchant wants to sell the goods at a price of $0.80M$, which is the selling price S.
> Thus, the profit is $S - 0.75L = 0.25S$. But, $S = 0.80M$.
> So, $0.80M - 0.25(0.80M) = 0.75L$.
> Or, $0.60M = 0.75L$.
> And therefore, $\frac{M}{L} = \frac{5}{4}$.

Applying the techniques, one can generate several new problems.

4.1.1 *Innovation by replacement*

> A merchant buys his goods at <u>20%</u> off the list price. He then marks the goods so that he can give his customers a discount of <u>10%</u> on the marked price but still make a profit of 25% on the selling price. What is the ratio of marked price to list price?

This problem is very similar to the original problem. Two of the given have been replaced (see underlined); nothing else was changed. The solution, therefore, differs only in quantity but not in the structure.

> Let L be the list price. The merchant bought the goods at the price of $0.80L$.
> Let M be the marked price. The merchant wants to sell the goods at a price of $0.90M$, which is the selling price S.
> Thus, the profit is $S - 0.80L = 0.25\text{S}$. But, $S = 0.90M$.
> So, $0.90M - 0.25(0.90M) = 0.80L$.
> Or, $0.675M = 0.80L$.
> And therefore, $\frac{M}{L} = \frac{32}{27}$.

4.1.2 *Innovation by addition*

> A merchant buys his goods at 25% off the list price. He then marks the goods so that he can give his customers a discount of 20% on the marked price but still make a profit of anywhere from 20% to 25% on the selling price. What is the ratio of marked price to list price?

The new problem generated by this innovation technique is quite different from the original problem. The added condition of a profit being anywhere between 20% to 25% makes the difference; the problem now involves an inequality. The solution is as follows:

> Let L be the list price. The merchant bought the goods at the price of $0.75L$.
> Let M be the marked price. The merchant wants to sell the goods at a price of $0.80M$, which is the selling price S.
> Thus, the profit is $S - 0.75L$. The desire is for this quantity to be anywhere from 20% to 25%.
> Thus, $0.20S \leq 0.80M - 0.75L \leq 0.25S$.
> But, $S = 0.80M$.
> So, $0.20(0.80M) \leq 0.80M - 0.75L \leq 0.25(0.80M)$.
> Solving the inequality gives $\frac{75}{64} \leq \frac{M}{L} \leq \frac{5}{4}$.

It is not much different; the inequality is really the only new feature in this problem.

4.1.3 *Innovation by modification*

> A merchant buys his goods at 25% off the list price. He then marks the goods so that he can give his customers a discount of 20% on the marked price but still make a profit of 25% on the selling price. If the merchant bought the goods at P200, what should be the marked price in order to realize said profit? [P represents pesos, the currency of the Philippines]

This problem is a slight modification of the original problem. If the original problem had not been solved, then this is not so easy because the student will still have to find a relationship between the list price and the marked price. But, if the original problem had been solved then this new problem is simple.

> To solve, one has to determine the list price L. The price of P200 is the net price after the discount. Thus, $L = \frac{200}{0.75} = 266.67$.
> From the original problem, the ratio of marked price to list price, given these same numbers is $\frac{5}{4}$.
> Therefore, $M = \frac{5}{4} \cdot 266.67 = 333.3\overline{3}$.

4.1.4 *Innovation by contextualizing the problem*

> Clara, who is into a book buy-and-sell business, buys her fiction books at 25% off the list price in a warehouse. She then marks the books so that when she sells these with a discount of 20% on the marked price she still makes a profit of 25% on the selling price. What could be a possible list price and marked price and how do these compare?

This new problem modernizes the context of the problem by talking about a particular person engaged in a particular business. Along with modernizing the context, a slightly different problem is posed, that of giving the marked price, when the list price is known and how these compare. The answer could still be a ratio of marked price and list price but a final answer would include a particular list price and marked price, slightly different from Problem c.

> Let L be the list price. Clara bought the books at the price of $0.75L$.
> Let M be the marked price. She wants to sell the goods at a price of $0.80M$, which is the selling price S.

Thus, the profit is $S - 0.75L = 0.25S$. But, $S = 0.80M$.
So, $0.80M - 0.25(0.80M) = 0.75L$.
Or, $0.60M = 0.75L$.
And therefore, $\frac{M}{L} = \frac{5}{4}$.

Therefore, if a book's list price is P100, then Clara should put down its marked price as P125. Or, if a book's list price was P240, Clara should mark this same book with the price of P300.

4.1.5 *Innovation by turning the problem around*

A merchant buys his goods at a discount. He then marks the goods so that he can also give his customers a certain discount on the marked price but still make a profit of 25% on the selling price. Suppose the merchant wants the marked price to be one and a half times the list price. What are the possible discount rates off the list price and the marked price?

This is not a very easy problem but the previous innovations and their solutions are a big help.

Let L be the list price, x be the decimal equivalent of the discount of the list price, M be the marked price, S be the selling price and y be the decimal equivalent of the discount off the marked price. Thus, $S - (1 - x)L = 0.25S$.

$$(1 - y)M - (1 - x)L = 0.25(1 - y)M$$

$$0.75(1 - y)\frac{3}{2}L - (1 - x)L = 0$$

$$\left[\frac{9}{8}(1 - y) - (1 - x)\right]L = 0$$

$$\frac{9}{8}y = x + \frac{1}{8}$$

$$y = \frac{8}{9}x + \frac{1}{9}.$$

Therefore, the two discount rates form a linear relationship, which is quite a revelation. This was due to the fixed relationship given for the marked price and the list price.

Thus, if L = 100, then M = 150. Suppose the discount on the list price is 25% or the merchant bought the item at 75. The discount on the marked price should be $y=\frac{8}{9}\cdot\frac{1}{4}+\frac{1}{9}=\frac{1}{3}$ or $33.\bar{3}\%$ in order to realize a fixed profit of 25% on the selling price $S = 100$.

As part of the solution, a table of possible values could also be made.

L	M	x	y	S	Profit on S	Profit Rate on S (%)
100	150	0.25	0.333333	100	25	25
100	150	0.2	0.288889	106.6667	26.666667	25
100	150	0.1	0.2	120	30	25
100	150	0.15	0.244444	113.3333	28.333333	25

4.1.6 *Innovation by reformulation*

> A merchant buys his goods at 25% off the list price. He then marks the goods so that he can give his customers a discount of 20% on the marked price but still make a profit of 25% on the selling price. Generalize the relationship between the marked price and the list price given a discount of X% on the list price, a discount of Y% on the marked price and a desired profit of Z% on the selling price.

The new problem generated by this technique requires a higher level of skill, that of generalizing a relationship between two quantities. This also requires a solid knowledge of variables.

> Let x, y, z be the decimal equivalent of X%, Y%, and Z%, respectively.
> Let L be the list price, M be the marked price, and S be the selling price. Then, the equation to be solved is $S-(1-x)L = zS$.
> But $S=(1-y)M$. Thus, $(1-y)M - z(1-y)M = (1-x)L$.

Solving, $(1-z)(1-y)M=(1-x)L$.

And, $\frac{M}{L}=\frac{1-x}{(1-z)(1-y)}$.

Thus, if we are to substitute, letting $x=0.25$, $y=0.20$, $z=0.25$, this is exactly the original problem. Then, $\frac{M}{L}=\frac{1-0.25}{(1-0.25)(1-0.20)}=\frac{5}{4}$, which is the answer to the original problem. Note the challenge of dealing with the many variables in order to solve this particular problem. But, clearly, too, this is generalizing the solution to the original problem.

Not all problems can be innovated on using all the techniques described above. For some problems, there are only one or two techniques that could be used for innovation.

Let us look at another example.

4.2 *Example 2*

The Problem:
Three cylindrical oil drums of 2-foot diameter are to be securely fastened in the form of a "triangle" by a steel band. What length of band will be required? (Krulik & Rudnick, 1989, p. 153)

To solve this problem, one must come up with the correct diagram:

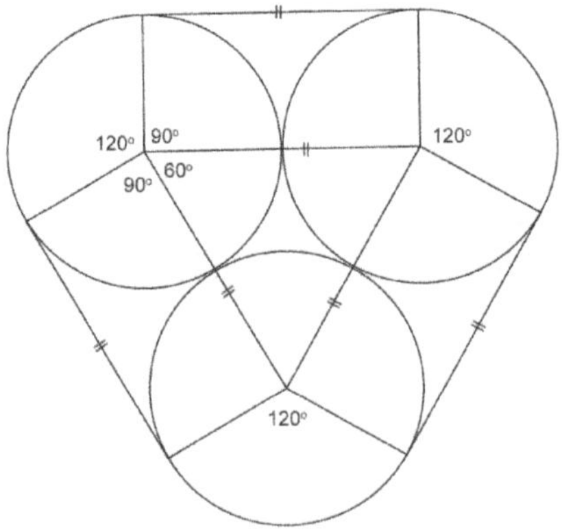

Figure 1. Krulik & Rudnick, 1989, p. 153

Based on Figure 1, the length of each rectangle is 2 feet and there are 3 rectangles. The lengths of the band around each circular drum make up the circumference of one drum, which is 2π. Therefore, the needed length of the band is $(6 + 2\pi)$ feet or approximately 12.28 ft.

4.2.1 *Innovation by replacement*

> Four cylindrical oil drums of 2-foot diameter are to be securely fastened in the form of a "square" by a steel band. What length of band will be required?

It is still helpful to have a diagram to work with.

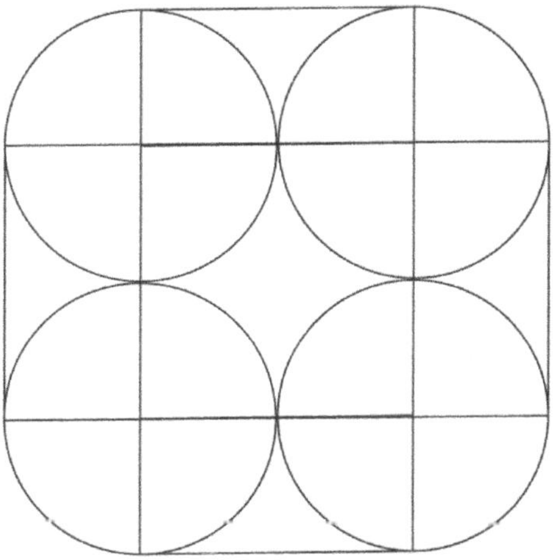

Figure 2. Four-cylinder problem

Once again, a correct diagram is important. Similar to the original problem, the length of each rectangle is 2 ft and therefore, the band length from one tangent point of the circle to the tangent point of the adjacent circle is 2 ft for a total of 8 ft. The bands around each circle, when put together make up the circumference of on circle with 2 ft diameter. Thus, the total length of the band needed for four drums is $(8 + 2\pi)$ ft or approximately 14.28 ft.

4.2.2 *Innovation by addition*

> Three cylindrical oil drums of 2-foot diameter are to be securely fastened in the form of a "triangle" by a steel band. A cylinder of 1-foot diameter is to be placed on the space between each pair of oil drums. All three of these cylinders are to be tied with the three drums. What length of band will be required?

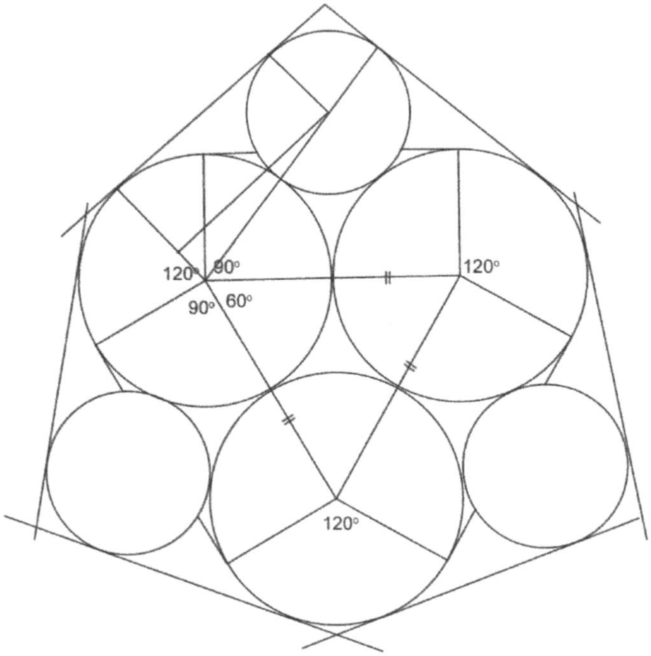

Figure 3. With three smaller cylinders

It can be shown that the length of one tangent from the small cylinder to the oil drum is $2\sqrt{2}$ ft because of the right triangle whose legs have lengths 3 ft and 1 ft. There are 6 of those tangents so the total length is $12\sqrt{2}$ ft. It can also be shown that the band wraps around $\frac{1}{6}$ of the circumference of each of the small cylinders and each of the oil drums. Each of the interior angles of the regular hexagon that circumscribes the cylinders and oil drums measures 120°, Therefore, the central angle of the sector wrapped by the band is 60°. The total curved length over the three cylinders is, in feet, $3\left(\frac{1}{6}\right)(2\pi)=\pi$. The total curved length over the three oil drums is, in feet, $3\left(\frac{1}{6}\right)(2\pi \cdot 2)=2\pi$. Therefore, the total length of the band needed is, in feet, $3\pi+2\sqrt{2} \approx$ 12.25.

4.2.3 *Innovation by modification*

> Three cylindrical oil drums of 2-foot diameter are to be securely fastened in the form of a "triangle" by a steel band. What is the total length of steel band that does not touch any of the drums?

This is slightly different. Using Figure 1, the problem can be easily answered. The total length is 6 ft.

4.2.4 *Innovation by contextualizing the problem*

> Bobby has to secure three pencils of the same size with transparent tape. Each pencil has a diameter of 1 cm. How long a tape does he need if he wants an overlap of 1 cm of the tape?

The problem gives a setting that students could relate to more. The solution is very similar to the original problem except that the diameter has been slightly changed to make the context more realistic. The total length is $(3 + \pi + 1)$ cm or approximately 7.14 cm.

4.2.5 *Innovation by turning the problem around*

> Suppose a packaging company has 15 feet of steel band available to fasten three cylindrical drums of 2-foot diameter each. Is this enough if they are to be fastened in the form of a triangle? If not, how much more does the company need? If yes, is there any left over? How much left over?

This is an obvious turnaround of the problem. It is a very good problem if the original problem had not been given yet. A fixed length for the steel band is given and using the same conditions as in the original problem, the new problem becomes, 'Is the length enough?'

Because the answer to the original problem is 12.28 ft, then 15 ft is enough with 2.72 ft left over.

4.2.6 *Innovation by reformulation*

> Three cylindrical oil drums of n-foot diameter are to be securely fastened in the form of a "triangle" by a steel band. Express the length of steel band needed in terms of n.

This is a generalization problem. Figure 1 is once more useful here. If the diameter is n-ft then the radius is $\frac{1}{2}n$. The straight length of the band will therefore total $3n$ ft. the curved lengths will be a total of, in feet, $2\pi \cdot \frac{1}{2}n$ or πn. Therefore, the length of steel band needed is, in feet, $3n + \pi n \approx 6.14n$.

5 Cognitive Value of Innovation in Problem Solving

I have cited that problem solving is a goal but the bigger picture of mathematics education indicates that problem solving is also a tool to deepen one's understanding and knowledge of mathematics. The ability to reason and prove is an indication that one has reached a very deep and complex understanding of mathematics. Indeed, reasoning and proving are two of the highest level of skills in the hierarchy of mathematical skills. This appears to be internationally accepted as it has been used in the TIMSS 2003 study (IEA, 2003). Problem solving certainly aims to develop those skills among students.

In his proposed analytic framework that focuses on developing reasoning and proving skills, Stylianides (2008) identified the following processes as comprising the activities of reasoning and proving:

"identifying patterns, making conjectures, providing non-proof arguments, and providing proofs" (p. 9). Further, he views the first two processes as "providing support to mathematical claims" and the last two as "making mathematical generalizations" (p. 9).

Taking cue from Stylianides (2008), results of problem innovation, as shown from the two examples, could certainly support the development of reasoning and proving skills among students. One valuable contribution of problem innovation is that some of the techniques generate more complex problems for students. The examples provided in this paper show that these complex problems are problems that require generalization and analysis of the structure of the problem. Both are higher–order skills as both indicate sophisticated levels of understanding of mathematical concepts (Stylianides, 2008). Furthermore, Arcavi and Resnick (2008) showed that not only could 'new' and complex problems come out of existing problems, 'new' and sophisticated solutions (e.g. geometric in addition to algebraic) to the original problem could also be produced even from a slight innovation of a problem and from the original solution itself. By continuously exploring a problem and its solution, students are able to generate more sophisticated ideas and reach a deeper level of mathematical knowledge and understanding.

6 Benefits of Innovation on Existing Problems

As pointed out in the earlier part of this paper, there does come a time when teachers run out of new problems to give to their class. By innovating on existing problems, teachers would be able to generate new problems, perhaps, even based on a favorite problem that they might have. Some old, classical problems are just too good to ignore. Using innovation, these old, classical problems can be given new forms. For some techniques, the 'new' problems may simply provide a new exercise for students to work on. But other techniques, when applied properly and carefully, could generate more complex problems that are useful in developing among students sophisticated problem solving techniques.

7 Practical Aspects of Innovation in Generating Mathematical Problems

Like any task of teaching, innovating on existing mathematical problems requires much thought and time. Innovation by *replacement* does not require much time or thought. Innovation by *addition* requires some thought but not much time. In using this technique, teachers need to make sure that the additional condition in the problem is sensible and reasonable. Innovation by *modification and reformulation* is quite tricky. One could easily interchange the two techniques, which is harmless. What is important is to know that these two techniques are meant to generate two different problems. Innovation by *contextualizing* could be a challenge because of the need to know what students could relate to in a particular period of time. Innovation by *turning the problem around* is clearly a challenge. This is because not all problems can be turned around to come up with a good, complex problem.

There are times when problems generated with the use of the innovation techniques are meaningless or do not have solutions. It is alright. The key is to try, explore, and try again until better problems are generated.

8 Concluding Remarks

Problem posing is a complex cognitive process (Silver et al, 1996). It is a skill that requires tremendous amount of work and practice. The innovation techniques discussed here provide mathematics teachers with ways to generate new problems and consequently, through problem exploration, help them develop better problem solving skills. By giving mathematics teachers more tools and techniques to increase their skills in any aspect of problem solving, we increase their confidence in this area of mathematics. The next step is to allow them time and opportunity to try out these techniques themselves.

Acknowledgement

The author wishes to thank Floredeliza F. Francisco, Ateneo de Manila University, The Philippines, for her invaluable insights and comments on the chapter.

References

Arcavi, A. & Resnick, Z. (2008). Generating problems from problems and solutions from solutions. *Mathematics Teacher, 102* (1), 10-14.

Branca, N. A. (1980). Problem solving as a goal, process, and basic skills. In S. Krulik & R. E. Reys (Eds.), *Problem solving in school mathematics* (pp. 3-8). Reston, VA: National Council of Teachers of Mathematics.

Brown, S. I. & Walter, M. I. (1983). *The art of problem posing.* NJ: Lawrence Erlbaum Associates.

Butts, T. (1980). Posing problems properly. In S. Krulik & R. E. Reys (Eds.), *Problem solving in school mathematics* (pp. 23-33). Reston, VA: National Council of Teachers of Mathematics.

Corbett, P. (2007). Telling tales. *English Teaching Online*, Spring (1) Term 3. Retrieved on January 23, 2008 from http://www.teachit.co.uk/custom_content/newsletters/newsletter_jan07.asp#1

Crespo, S. (2003). Learning to pose mathematical problems: Exploring changes in preservice teachers' practices. *Educational Studies in Mathematics, 52* (3), 243-270.

Crespo, S. & Sinclair, N. (2008). What makes a problem mathematically interesting? Inviting prospective teachers to pose better problems. *Journal of Mathematics Teacher Education 11*, 395-415.

Garces, I. J. (2008). MTAP *Individual competition secondary level.* Unpublished document.

IEA (2003). TIMSS *assessment frameworks and specifications.* Boston: IEA.

Krulik, S. & Rudnick, J. A. (1989). *Problem solving: A handbook for senior high school teachers.* Boston, MA: Allyn and Bacon.

Malone, J. A., Douglas, G. A., Kissane, B. V. & Mortlock, R. S. (1980). Measuring problem solving ability. In S. Krulik & R. E. Reys (Eds.), *Problem solving in school mathematics* (pp. 204-216). Reston, VA: National Council of Teacher of Mathematics.

National Council of Teachers of Mathematics (NCTM). (2000). *Principles and standards for school mathematics.* Reston, VA: NCTM.

Polya, G. (1945). *How to solve it.* Princeton, N. J.: Princeton University Press.

Schoen, H. L. & Oehmke, T. (1980). A new approach to the measurement of problem-solving skills. In S. Krulik & R. E. Reys (Eds.), *Problem solving in school mathematics* (pp. 216-227). Reston, VA: National Council of Teachers of Mathematics.

Singapore Ministry of Education. (2006b). *Primary mathematics syllabus.* Singapore: Curriculum Planning and Development Division.

Singapore Ministry of Education. (2006b). *Secondary mathematics syllabus.* Singapore: Curriculum Planning and Development Division.

Silver, E. A., Mamona-Downs, J., Leung, S. S. & Kenney, P. A. (1996). Posing mathematical problems: An exploratory study. *Journal for Research in Mathematics Education, 27* (3), 293-309.

Stylianides, G. J. (2008). An analytic framework of reasoning-and-proving. *For the Learning of Mathematics, 28* (1), 9-16.

Teo, K. M., To, W. K. & Wong, Y. L. (2000). *Singapore secondary school mathematical olympiads 1999 - 2000*. Singapore: Singapore Mathematical Society.

Wong, K. Y. (Ed.) (1996). *New elementary mathematics (Syllabus D)*. Singapore: Pan Pacific Publication.

Yeap, B. & Kaur, B. (1999). Mathematical problem posing: An exploratory investigation. In E. B. Ogena & E. F. Golla (Eds.), *Mathematics for the 21st Century, 8th Southeast Asian Conference on Mathematics Education Technical Papers* (pp. 77-86). Taguig, Philippines: SEAMS & MATHTED.

Appendix

Use any of the innovation techniques to generate new problems out of the following:

1. How many different triangles with integer sides can be drawn having the longest side (or sides) of length 6? How many of the triangles are isosceles? (Butts, 1980, p. 25)

2. What values are possible for the area of quadrilateral *EKDL* if *ABCD* and *EFGH* are squares of side 12 and E is the center of square *ABCD*? (Butts, 1980, p. 30)

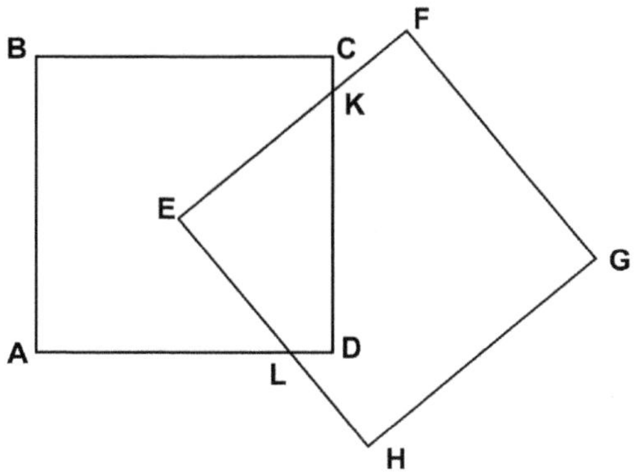

3. A 6-foot tall man looks at the top of a flagpole making an angle of 40° with the horizontal. The man stands 50 feet from the base of the flagpole. How high is the flagpole to the nearest foot? (Krulik and Rudnick, 1989, p. 161)

4. A piece of "string art" is made by connecting nails that are evenly spaced on the vertical axis to nails that are evenly spaced on the horizontal axis, using colored strings. The same number of nails must be on each axis. Connect the nail farthest from the origin on one axis to the nail closest to the origin on the other axis. Continue in this manner until all nails are connected. How many intersections are there if you use 8 nails on each axis? (Krulik and Rudnick, 1980, p. 164)

5. A figure is divided into five regions as shown in the diagram below. Given 4 distinct colors, how many different ways are there to colour the figure so that no two regions with a common boundary receive the same colour? (Teo, To and Wong, 2000, p. 21)

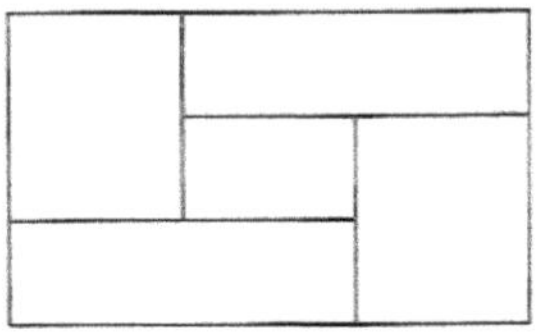

6. Eight points on a circle are grouped into disjoint pairs. Each pair is joined by a chord. Find the number of ways of joining pairs such that no two of the chords intersect. (Teo, To and Wong, 2000, p. 20)

7. A 6 x 7 rectangle is divided into 6 x 7 unit squares as shown. What is the total number of squares of all sizes in the rectangle? (Wong, 1996, p. 37)

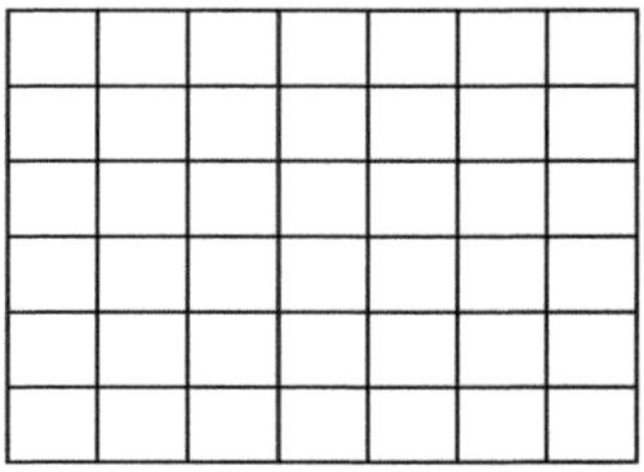

8. In 1991, a family spent 19% of their income on rent, 26% on food, 30% on other items and saved the rest. In 1992, their income increased by 10%. If the cost of food increased by 10%, savings decreased by 4% and rent remained the same, by what percentage did the expenditure on other items increase? (Wong, 1996, p. 211)

9. When a mother was 3 times as old as her son was, she was as old as he is now. When the son is as old as his mother is now, she will be 70 years old. How old is the mother now? (Wong, 1996, p. 360)

10. Two swimmers at opposite ends of a 90-foot pool, start to swim the length of the pool, one at the rate of 3 feet per second, the other at 2 feet per second. They swim back and forth for 12 minutes. Allowing no loss of time at the turns, find the number of times they pass each other. (Garces, 2008)

Chapter 11

Mathematical Problems for the Secondary Classroom

Jaguthsing DINDYAL

Problems abound in mathematics education at all levels. This chapter focuses on some of the desirable skills that secondary level teachers can develop among students through the use of selected problems. Amongst others, we wish students to develop the following skills while solving problems: generalising and extending problems; using different representations to solve problems; making connections between different content areas; using technology in significant ways; drawing or constructing; proving and explaining; carrying out simple investigations; formulating problems; and solving open-ended problems. Each type of skill is highlighted with an example followed by a brief comment.

1 Introduction

A typical mathematics textbook for any level is full of so-called "mathematical problems". A typical student learning mathematics at any level has to solve many mathematical problems. The typical mathematics teacher has to either write new problems or find relevant problems from reliable sources for the students to learn mathematics while doing these problems. Thus, we find that mathematical problems are inherent in the structure of the subject itself. Problems have a long history and have occupied a central place in the school mathematics curriculum since antiquity, although the same cannot be said about problem solving (Stanic & Kilpatrick, 1988). From the perspective of a teacher, two aspects of mathematical problems stand out:

1. writing original problems, modifying standard problems or sourcing for relevant problems, and
2. using different problems effectively in class to develop desirable skills among mathematics students.

In this paper, I shall look partly into the second aspect. I shall first comment on what constitutes a problem and then focus on some of the desirable skills that we can develop among students through the use of problems (the term skill will be used a little loosely in the context of this paper). Any discussion about problems cannot be divorced from the problem solving process. But what is a problem and what is problem solving?

2 Problem and Problem Solving

If the answer to a "problem" is apparent then it is no longer a "problem". Hence, the defining feature of a problem situation is that there must be some blockage on the part of the potential problem solver (Kroll & Miller, 1993). A problem can be considered as a *task* which elicits some activity on the part of students and through which they learn mathematics during the problem solving activity. One of such descriptions is by Lester (1983) who defined a problem as a task for which:

1. the individual or group confronting it wants or needs to find a solution;
2. there is not a readily accessible procedure that guarantees or completely determines the solution; and
3. the individual or group must make an attempt to find a solution (p. 231-232).

It is interesting to note that Lester's definition accommodates not only the individual's perception but also the group's perception on what constitutes a problem. Along similar lines, Krulik and Rudnik (1980) defined a problem as a situation that requires resolution and for which the individual sees no apparent or obvious means or path to obtaining the solution. It should be reiterated that the solver or solvers should be motivated to reach the solution because what is a problem for one person may not necessarily be a problem for another person. Schoenfeld (1985) clearly pointed to the difficulty in describing what constitutes a problem:

> The difficulty with the word problem is that problem solving is relative. The same task that calls for significant efforts from some students may well be routine exercises for others, and answering them may just be a matter of recall for a given mathematician. Thus being a "problem" is not a property inherent in a mathematical task. Rather, it is a particular relationship between the individual and the task that makes the task a problem for that person (p. 74).

Although Schoenfeld seems to focus on an individual's relationship to the problem, it would be unfair to say that he excludes problems which require any collaborative work.

Polya (1957, 1966) differentiated between routine and non-routine problems. While routine problems are mere exercises that can be solved by some rules or algorithms, non-routine problems are more challenging, and they require some degree of creativity and originality from the solver. Polya added that it is only through the judicious use of non-routine problems that students can develop problem solving ability. Accordingly, problem solving is not just about solving a problem. It is the process by which students experience the power and usefulness of mathematics in the world around them and it also a method of inquiry and application (National Council of Teachers of Mathematics [NCTM], 1989). Thus, problem solving is a complex process which Polya (1957) claimed proceeds through his much publicized four phases: *understanding the problem*, *devising a plan*, *carrying out the plan*, and *looking back*. Problem solving has been used with multiple meanings that range from "working rote exercises" to "doing mathematics as a professional" (Schoenfeld, 1992).

In 1980, the publication of the *Agenda for Action* by the NCTM in the United States, spurred new interest in problem solving. The statement that problem solving should become the focus of school mathematics was widely publicized. One of the goals set by the NCTM (1989) for K-12 mathematics education was that students become mathematical problem solvers, if they were to become productive citizens. It was perceived as important that students solve problems not only alone, but also working cooperatively in small or large groups. Problems had to be varied, with some being open-ended and in more applied contexts. Regarding the grades 5-8 curriculum, the NCTM (1989) standards stated that it should take advantage of the expanding mathematical capabilities of middle school students to include more complex problem situations involving topics such as probability, statistics, geometry, and rational numbers. The standards also suggested that some of the problems had to be more demanding, requiring extended effort from the students. It was also claimed that students had to make full use of available technology as problem-solving tools and that they had to learn to work cooperatively on selected problems.

The NCTM (2000) standards further acknowledged the important role of problem solving in mathematics education at school level. Problem solving is highlighted as one of the five process standards that cut across the curriculum at all grade levels. The problem solving standard states that instructional programs at school level should enable students to: build new mathematical knowledge through problem solving; solve problems that arise in mathematics and in other contexts; apply and adapt a variety of appropriate strategies to solve problems; monitor and reflect on the process of problem solving. The standards also mentioned that students had to reflect on their problem solving and consider how it might be modified, elaborated, streamlined, or clarified.

Problems in the school mathematics curriculum have changed significantly over time depending on what has been emphasised during those times. Several factors can be identified that differentiate one problem from another. Amongst others, problems differ by: the content domain, the objectives to be tested, the exact wording of the problem, the context of the problem, the support and structure provided, the types of numbers involved, the resources to be used during the solution process, the expected time for a solution, and the *closedness* or *openness* of the problem.

Since we wish to teach highly desirable mathematical concepts and skills, the problem tasks that we choose must meet certain criteria. The NCTM (1991) claimed that good tasks are ones that do not separate mathematical thinking from mathematical concepts or skills, they capture students' curiosity, and they invite the students to speculate and to pursue their hunches. However, it is important to note that there are some distinctions associated with the tasks at various levels (Mason & Johnston-Wilder, 2006, p. 8):

- the task as imagined by the task author;
- the task as intended by the teacher;
- the task as specified by the teacher-author instructions;
- the task as construed by the learners; and
- the task as carried out by the learners.

Accordingly, whether a problem task is originally written down by a teacher or is taken or modified from a secondary source, it carries an implicit intent that the assigner of the task (in this case the teacher) wishes to achieve by assigning the problem task to the solver. There are bound to be mismatches between what the assigner wishes to achieve and what actually is achieved during the solving process. Thus, problem tasks when used in the classroom have many such underlying nuances that need to be considered.

Amongst others, we wish students to develop the following skills while solving problems: generalising and extending problems; using different representations to solve problems; making connections between different content areas; using technology in significant ways; drawing or constructing; proving and explaining; carrying out simple investigations; formulating problems; and solving open-ended problems. Some problems which can elicit the desirable knowledge and skills are described below.

2.1 *Using generalisation and extension*

Kaput (1999) claimed that generalization involves deliberately extending the range of reasoning or communication beyond the case or cases considered by explicitly identifying and exposing commonality across the case or the cases. He added that this resulted in lifting the reasoning or communication to a level where the focus is no longer on the cases or situations themselves but rather on the patterns, procedures, structures, and relations across and among them, which in turn become new, higher-level objects of reasoning or communication. Consider the following problem:

How many squares are there on a regular chess board?

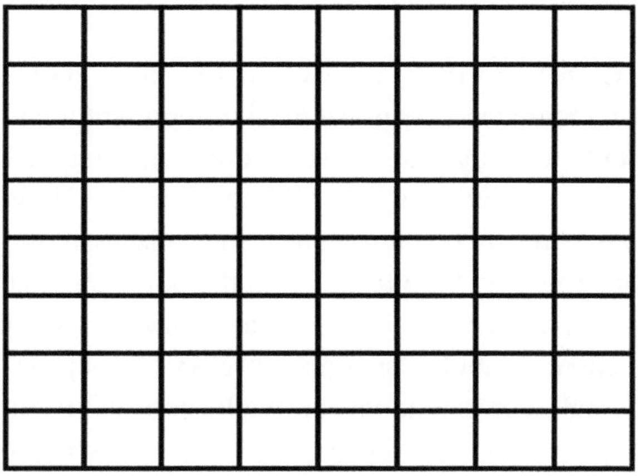

The problem can be asked with or without the visual support showing the 8×8 grid. The students can develop various skills through the process of solving this problem. One of the heuristics to be used includes considering simpler cases. The students can try a 1×1 grid, a 2×2 grid and a 3×3 grid to generate a pattern. In this inductive approach the student can find the solution that for a regular 8×8 chess board, we have $1 + 4 + 9 + \ldots + 64 = 204$ squares. In the problem solving process, we cannot just stop at this point. Besides checking the solution, teachers can encourage students to generalise and extend the problem. A simple generalisation of the problem would be to find the number of squares in an $n \times n$ grid. It is expected that students will be able to write without much difficulty, the answer as $1^2 + 2^2 + 3^2 + \ldots. + n^2$. Although, it is not required that students at this level know about $\sum r^2 = \frac{1}{6}n(n+1)(2n+1)$, the teacher may consider exploring this idea further, depending on the ability level of the students and the time that is available. Furthermore, the teacher may consider an extension of the problem to find the number of rectangles instead of the number of squares in an $n \times n$ grid. Although, students may use a similar approach as for the number of squares, they may meet with some difficulties. A further extension could be to find the number rectangles in an $m \times n$ grid. Students can learn much more mathematics by solving these extended problems.

2.2 *Using different representations*

Hiebert and Carpenter (1992) claimed that to think about mathematical ideas and to communicate them, we need to represent them in some way. Thinking about mathematical ideas requires an internal representation that allows the mind to operate on them. However, mental representations are not observable and discussions about these representations can only be inferred. On the other hand, communication requires that the representations be external, taking the form of spoken language, written symbols, pictures or physical objects. Connections between external representations of mathematical information can be constructed by the learner between different representational forms of the same mathematical idea or between related ideas of the same representational form. Hiebert and Carpenter also added that there is an ongoing debate whether mental representations mimic in some way the external object being represented or whether there is a common form used to represent all information. As such, students must have the exposure to different forms of representations while solving problems. They must not only understand the symbolic, numerical and graphical forms of representations but must also be able to move flexibly in between these forms of representations. While some problems emphasise one particular form of representation, others can be solved using various representations. Consider the following problem:

> *The Nice car rental agency charges $70 a day and 40 cents per kilometre. The Good car agency charges $60 a day and 50 cents a kilometre. Which agency will you choose to rent a car for a day? Give reasons for your answer.*

This problem can be solved by using all three forms of representations: by using algebra, or using a table of values or by using graphs.

(a) Using algebra

Let x be the number of kilometres travelled in a day.

Daily cost of a car from Nice car rental agency = $\$(70 + 0.4x)$

Daily cost of a car from Good car rental agency = $\$(60 + 0.5x)$

When $\$(70 + 0.4x) = \$(60 + 0.5x)$, $x = 100$ km.

As such, for distances less than 100 km, Good car rental agency is better. However, for distances greater than 100 km, Nice car rental agency is better.

(b) Using a table of values

Distance (km)	20	40	60	80	100	120	140
Nice rental agency ($)	78	86	94	102	110	118	126
Good rental agency ($)	70	80	90	100	110	120	130

The table clearly shows that for distances less than 100 km, Good rental agency is better whereas for distances greater than 100 km, Nice rental agency is better.

(c) Using a graph

$C = 70 + 0.4x$ for Nice car rental agency

$C = 60 + 0.5x$ for Good car rental agency

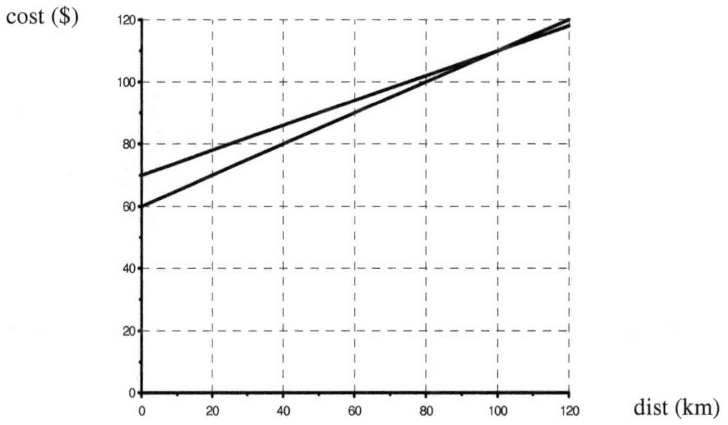

The graph shows that when the distance travelled is less than 100 km then Good car rental agency is better whereas Nice car rental agency is better for distances of 100 km or more.

Teachers can let the students struggle with the problem and choose a method that is more familiar to them. After which, the teacher may ask for alternative ways of solving the problem. The teacher can then highlight the fact that this problem can be solved using three different representations.

2.3 *Making connections*

Connections are problem solving tools and the teacher's task is to promote the use of connections in problem solving (Hodgson, 1995). When students can see the connections across content areas, they develop a view of mathematics as an integrated whole (NCTM, 2000). Carefully chosen problems can help students to make connections. Consider this problem:

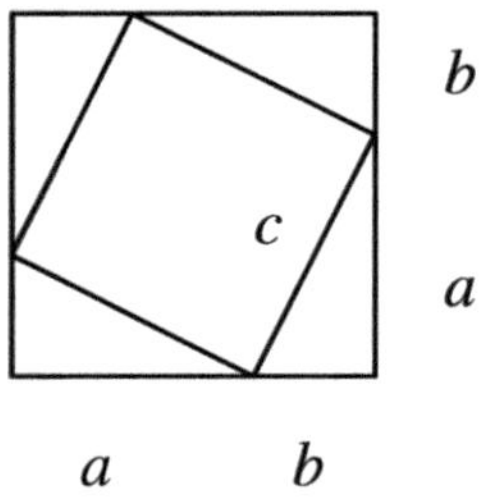

Four right angled triangles, each having shorter sides of lengths a and b and hypotenuse of length c, are joined together to form the figure as shown in the diagram above. Explain why the area enclosed by the four triangles is a square. How can you use the above figure to prove Pythagoras Theorem?

The interplay of algebra and geometry is obvious in this problem which requires the students to show that $(a + b)^2 = 4 \times \frac{1}{2}ab + c^2$ and hence deduce Pythagoras Theorem. Teachers need to highlight this important connection between the two branches of mathematics. Consider this simple integration problem: $\int_{-1}^{3} |x-1| dx$. The student, who goes about using the signs of $|x-1|$ on the given interval, may reach a solution. However, the student who can make the connection between the definite integral and an area under the graph certainly demonstrates a deeper understanding of the mathematical concepts involved. Teachers can make good use of such problems to make connections between important ideas in mathematics.

2.4 *Finding multiple methods of solution*

Students must be exposed to problems which can be solved in various ways. Even after solving a problem, students should be encouraged to look for *alternative solutions*, which is an important step in the Polya's model (see Polya, 1957).

Given that the angle sum of all pentagrams or five-cornered stars is constant, determine that angle sum. Use as many different methods as you can.

There are several ways in which this problem can be solved (see Lipp, 2000). Students learn by solving a problem in several ways. They also come to understand why some solutions are more elegant than others. Teachers can let students try their own methods and discuss if there are other possible methods of solution or a teacher may also ask students: "I want you to find at least two different ways of solving this problem." This may create an interest in the multiple ways of solving the problem.

2.5 *Using technology*

Using technological tools, students can reason about more-general issues and can model and solve complex problems that were heretofore inaccessible to them (NCTM, 2000). Technology can not only be used to help students do routine procedures such as calculations but also to help them in modelling and simulation activities. For example, to solve the following problem, students can use simulation to obtain an estimate of the probability:

Out of any group of five people what do you think is the chance that at least two of them will have a birthday in the same month?

On spreadsheet, students can generate random samples of 5 numbers from 1 to 12 to get an estimate of the probability. The same can be done using a graphing calculator as well. The simulation exercise can provide important insights into how to solve the actual problem itself. Consider the following problem:

Show that the midpoints of the sides of a quadrilateral form a parallelogram.

Students may not quite have an idea about how to work out the solution. However, if they use dynamic geometry software such as the Geometer's Sketchpad (GSP) students will benefit from the exploration in arriving at a solution. Other graphing software can help students with mathematical problems involving graphs. Teachers should carefully select problems so that at least a few may require some use of technology in their solution.

2.6 *Drawing or constructing*

Students must be given ample opportunities to draw or construct. These constructions must not be routine ones; rather, the students need to demonstrate ample understanding of the conceptual underpinnings. For example, students need to mobilise their geometrical knowledge in order to solve the problem below:

Construct a triangle with the same area as the trapezium shown below.

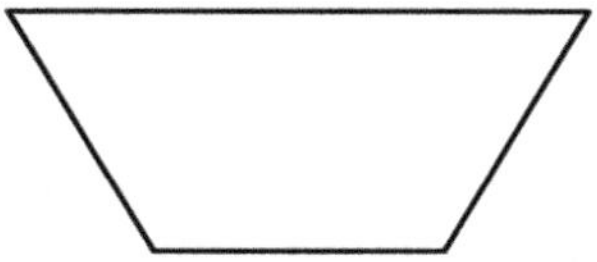

Although at first sight, this problem seems easy, it is quite demanding on the average student. A student doing the construction will need to know how to manipulate the geometrical tools and how to draw triangles having the same area. There are several ways in which the construction can be carried out. Teachers can emphasise the use of the drawing instruments. In this case, it is important for students to know that points on a line parallel to the base of a triangle form other triangles with that base and have the same area as the given triangle.

Consider this new problem:

A student notices what appears to be an arc of a circle. How can she justify that this figure is really part of a circle?

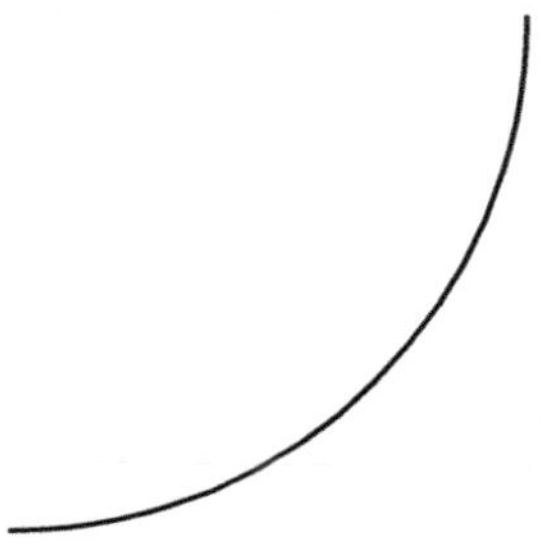

A student who wishes to solve the problem will need to remember how to construct the centre of a circle given an arc of the circle. Such construction problems require students to know about the underlying geometrical principles. In this case, it is important for the students to know that the perpendicular bisector of a chord of a circle passes through the centre of the circle.

2.7 *Proving*

Proof is an important part of mathematics and in the classroom the key role of proof is the promotion of mathematical understanding (Hanna, 2000). However, not all students are able to do proofs and proving does not come naturally to them. The students need to be exposed a wide variety of problems which requires them to prove. The simplest results that often students take for granted may be the best starting points. For example:

> *If ABC is an isosceles triangle, prove that it has two congruent angles.*

Many students assume the congruent angles are an obvious aspect of the triangle being isosceles and there is no need to prove. Teachers need to carefully detail the elements of a proof and help students to write their statements supported by strong reasons. Teachers may, for example, point to: What is given? What is to be proved? Do you need to draw a diagram?

Many students do not feel the need to prove that the sum of the interior angles in a triangle is 180°. Other proof problems may include:

Prove the following properties related to circles:

a. *Equal chords are equidistant from the centre.*
b. *The perpendicular bisector of a chord passes through the centre.*
c. *Tangents from an external point are equal in length.*

Prove the midpoint theorem: A straight line joining the mid-

points of two sides of a triangle is parallel to the third side and is equal to half of it.

Proof should not be limited to the topic on geometry. Students should see proof as a natural aspect of mathematics. Other simple proofs can include:

Prove that if n is even then n^2 is even.
Prove that if n is odd then n^2 is odd.

Even selected proofs of identities from trigonometry can be used. It is the teacher's role to select such proof items from various topics and let students have ample practice with the proofs.

2.8 *Carrying out simple investigations*

Students should be given the opportunities to carry out simple investigations. In the real world, these types of investigations help us to solve problems.

One way to make a rectangular container is to take a rectangle of Vanguard paper and cut the same size square out of each corner and then fold the four sides up and tape the corners.
If you began with a piece of paper 20 cm by 15 cm, what size square should you cut out of each corner to make the volume of the box as large as possible?
What will be the dimensions of the resulting box if you start with a piece of A4 paper?

The investigations should not be too difficult for the students, but should be matched to their level of mathematical understanding. Investigation problems have the added advantage of encouraging students to make joint efforts by working in small groups. Working collaboratively on problems is one of the skills we wish students to develop among students. Statistics is a good topic for giving investigative type of problems to students.

2.9 *Solving open-ended problems*

Students get so used to solving problems that have only one solution that they cannot think about problems with more than one solution. This conditioning may have serious consequences for students when solving real-life problems. The following problems are open-ended:

A number rounded to 2 decimal places is 2.34. What could be the number?

After five games, the CLAM football club has averaged 3 goals per game. What might have been its scores in each of the five games?

The probability that both Jane and Bill go to school by bus is 0.03 on a particular day. What could be the probability of each one of them separately going to school by bus on that day?

$\int_1^4 f(x)\mathrm{d}x = 8$, *find* $f(x)$.

The above problems make the students think deeply about the underlying mathematical concepts. Their answer cannot be just a recall of previously learned facts or skills.

Sullivan and Clarke (1991) used the idea of *Good Questions* to refer to such problems. These authors claimed that good questions (1) require more than the recall of a fact or reproduction of a skill, (2) pupils learn by doing the task, and the teacher learns about the pupil from the attempt, and (3) there may be several acceptable answers. Teachers should let students practice solving open-ended problems. Open-ended problems are not difficult to create. For example, instead of asking: *what is the area of a rectangle with sides 3 m and 4 m?*, one may ask: *find the sides of a rectangle with area 12* m^2.

2.10 *Formulating a problem or problem posing*

Another aspect of problem solving that is seldom included in textbooks is problem posing, or problem formulation (Wilson, Fernandez, &

Hadaway, 1993). The authors added that problem posing and problem formulation are logically appealing notions to mathematics educators and teachers. In the classroom, students are typically asked to solve problems rather than formulate or pose problems. A highly desirable skill which all students should possess is the ability to formulate their own problems. Problem formulation is not a routine exercise. It involves deep thinking about the given information and how to use that information to generate or formulate a problem having an idea about how to solve the problem. For example:

> *The following items are for sale:*
> *Pencil for $0.60*
> *Ruler for $1.00*
> *Copybook for $1.50*
> *A poster for $0.90*
> *A girl has four one-dollar coins, three 50-cent coins, four 20-cent coins and three 10-cent coins.*
> *Formulate a problem using the information given above.*

> *The line L has equation* $y = 2 - x$ *and the curve C has equation* $y = x^2$. *Formulate a problem using the above information.*

It is expected that the students will think deeply about the mathematical concepts and principles involved before formulating a problem. However, problem posing is not only about starting with some facts and generating a problem. Problem posing can also accommodate changing given problems or generating new problem from existing ones by changing certain conditions. For example, extending a given problem involves an aspect of problem posing whereby some conditions are changed. In the initial stages, teachers should help students to formulate problems and explain to them what problem formulation means. Gradually, this should become a natural process in the classroom.

3 Conclusion

The skills highlighted above are not trivial for any student learning mathematics. These skills have to be nurtured over a long period of time through the use of carefully selected problems by the mathematics teacher. A single problem cannot bring out all of these desirable skills; rather we need to use a range of problems spread over various topic areas. At the secondary level, as students are exposed to more and more mathematics, they should be gradually led to solve harder and harder problems which test them on the desirable skills we wish the students to have. The above list and type of problems is by no way exhaustive. However, the aim in this paper was to provide an overview of some types of mathematical problems that teachers might consider while teaching mathematics at the secondary level to develop some desirable skills.

References

Hanna, G. (2000). Proof, explanation and exploration: An overview. *Educational Studies in Mathematics, 44*(1-2), 5-23.

Hiebert, J., & Carpenter, T. P. (1992). Learning and teaching with understanding. In D. A. Grouws (Ed.), *Handbook of research on mathematics teaching and learning* (pp. 65-97). NewYork: Macmillan.

Hodgson, T. (1995). Connections as problem solving tools. In P. A. House & A. F. Coxford (Eds.), *Connecting mathematics across the curriculum: 1995 NCTM Yearbook* (pp. 13-21). Reston, VA: National Council of Teachers of Mathematics.

Kroll, D. L., & Miller, T. (1993). Insights from research on mathematical problem solving in the middle grades. In D. T. Owens (Ed.), *Research ideas for the classroom: Middle grades mathematics* (pp. 58-77). New York: Macmillan Publishing Company.

Kaput, J. J. (1999). Teaching and learning a new algebra. In E. Fennema, & T. Romberg (Eds.), *Mathematics classrooms that promote understanding* (pp. 133-155). Mahwah, NJ: Lawrence Erlbaum Associates.

Krulik, S., & Rudnik, J. A. (1980). *Problem solving: A handbook for teachers*. Boston: Allyn & Bacon.

Lester, F. K. (1983). Trends and issues in mathematical problem-solving research. In R. Lesh & M. Landau (Eds.), *Acquisition of mathematics concepts and processes* (pp. 229-261). Orlando, FL: Academic Press.

Lipp, A. (2000). The angles of a star. *Mathematics Teacher, 93*(6), 512-516.

Mason, J., & Johnston-Wilder, S. (2006). *Designing and using mathematical tasks*. St Albans: Tarquin Publications.

National Council of Teachers of Mathematics. (1989). *Curriculum and evaluation standards for school mathematics*. Reston, VA: Author.

National Council of Teachers of Mathematics. (1991). *Professional standards for teaching mathematics*. Reston, VA: Author.

National Council of Teachers of Mathematics. (2000). *Principles and standards for school mathematics*. Reston, VA: Author.

Polya G. (1957). *How to solve it?* (2nd ed.). New York: Doubleday & Co.

Poly, G. (1966). On teaching problem solving. In E. G. Begle (Ed.), *The role of axiomatics and problem solving in mathematics* (pp. 123-129). Boston, MA: Ginn and Company.

Schoenfeld, A. H. (1985). *Mathematical problem solving*. Orlando, FL: Academic Press Inc.

Schoenfeld, A. H. (1992). Learning to think mathematically: Problem solving, metacognition, and sense making in mathematics. In D. A. Grouws (Ed.), *Handbook of research on mathematics teaching and learning* (pp. 334-370). New York: Macmillan Publishing Company.

Stanic, G. M. A., & Kilpatrick, J. (1988). Historical perspectives on problem solving in the mathematics curriculum. In R. I. Charles, & E. A. Silver (Eds.), *The teaching and assessing of problem solving* (pp. 1-22). Reston, VA: National Council of Teachers of Mathematics.

Sullivan, P., & Clarke, D. (1991). *Communication in the classroom: The importance of good questioning*. Geelong, Vic, Australia: Deakin University Press.

Wilson, J. W., Fernandez, M. L., & Hadaway, N. (1993). Mathematical problem solving. In P. S. Wilson (Ed.), *Research ideas* for *the classroom: High school mathematics* (pp. 57-78). Reston, VA: National Council of Teachers of Mathematics.

Chapter 12

Integrating Open-Ended Problems in the Lower Secondary Mathematics Lessons

YEO Kai Kow Joseph

This chapter describes the characteristics of open-ended problems and the processes involved in solving such problems at the lower secondary (grades 7 and 8) level. Four examples of open-ended problems are used to demonstrate the benefits of integrating open-ended problems into mathematics lessons.

1 Introduction

The revised framework of the Singapore mathematics curriculum continues to encompass mathematical problem solving as its central focus. There are some changes in the framework components. For example, 'reasoning, communication and connections' and 'applications and modelling' are now included as processes that should receive increased attention (Ministry of Education, 2006). This is in line with similar reform-based visions of schooling around the world (National Council of Teachers of Mathematics, 2000; NSW Board of Studies, 2002). It is evident from the curriculum framework that problems are both a means as well as an end. While the primary purpose of teaching mathematics in Singapore schools is to enable students to solve problems, mathematics is also viewed as an excellent vehicle for the development and improvement of students' intellectual ability. Since problem solving was made the focus of the curriculum in the 1990s, teachers have been encouraged to cover a wide range of problem situations from routine mathematical problems to problems in an unfamiliar contexts and open-ended investigations (Ministry of Education, 1990, 2000). Integrating open-ended problems into mathematics lessons remains a challenge for many mathematics teachers.

Most of the problems found in earlier Singapore's mathematics textbooks intended for lower secondary (grades 7 and 8) students were routine and closed (Fan & Zhu, 2000). Research has shown that most students' experiences in schools focus on well-defined problems (Eggen & Kauchak, 2001). Such well-defined problems expect students to be able to apply and practise recently-acquired algorithms (Kulm, 1994). The situation is aggravated by the fact that teachers tend to 'load' students with rules, algorithms and formulae as they would with machines. Students are expected to commit to memory, and to be able to regurgitate, formulae, as well as to solve such well-defined problems faultlessly. Not only has this resulted in the students listening and absorbing knowledge passively, but it has also led students into developing mathematics avoidance (Collin, Brown & Newman, 1989). It is, therefore, heartening to note that recent lower secondary school mathematics textbooks series in Singapore have introduced many new types of problems. In particular, a few open-ended problems are found at the end of almost all chapters in several textbooks series (Chow & Ng, 2007, 2008; Lee & Fan, 2007, 2008; Sin & Chip, 2007, 2008).

In my work with lower secondary mathematics teachers in Singapore, I frequently hear the concerns about integrating open-ended problems into the mathematics curriculum. In addition, lower secondary mathematics teachers may feel inadequate about their own teaching approaches to problem solving, especially with open-ended problems. There is a need to equip lower secondary mathematics teachers with a set of greater variety of mathematical open-ended problems to enhance their teaching techniques. Students gain in many ways while solving open-ended problems. Students benefit because they need to make decisions and plan strategies as well as to apply their mathematical knowledge to the open-ended problems. The main purpose of this chapter is to review the characteristics of open-ended problems and the process of solving such problems. This chapter includes four open-ended problems that can be integrated into the teaching and learning of mathematics at the lower secondary level. When integrating open-ended problems in mathematics lessons, teachers need to focus less on the final answer and more on the thinking and concepts so that students see the value of mathematics.

2 Characteristics of Open-Ended Problems

Various researchers have their own views as to what constitutes an open-ended problem. Although there is no universal definition of an open-ended problem, we can still identify some of its basic characteristics. Open-ended problems are often considered as tasks in which there are more than one correct solution. Such problems allow students to use many approaches to solve them by placing few restrictions on students' solution methods (Hancock, 1995). According to Becker and Shimada (1997), when students are asked (1) to find several or many correct answers, (2) to find several or many different correct approaches to get an answer, or (3) to formulate or pose problems of their own, the students are said to be solving an open-ended problem. They further emphasize that the 'openness' of a problem is lost if the teacher proceeds as though there is only one correct answer or one method is presupposed to be the correct one.

In Singapore, Foong (2002) broadly classified problems as "closed or open-ended in structure" (p. 18). She elaborated that closed problems were 'well structured' in terms of clearly formulated tasks where one correct answer could be found in a fixed number of ways from the necessary data given in the problem setting. She further stated that open-ended problems were deemed as 'ill-structured' as they lack clear formulation. Such open-ended problems may have missing data or require assumptions and there is no fixed process that can guarantee a correct answer. Many real-world problems fall under this open-ended category. Such problems are set in contexts and are helpful for students to appreciate the real-life significance of mathematical concepts. Moreover, such problems also have the added benefit of helping students grasp concepts through linking abstract, unfamiliar mathematical concepts to real-life situations.

Furthermore, Sullivan and Lilburn (2005) expressed that open-ended problems are exemplars of good questions in that they advance significantly beyond the surface. Specifically, they also indicated that open-ended problems are those that require students to think more intensely and to provide a solution which involves more than remembering a fact or repeating a skill. Meanwhile, Leatham, Lawrence and Mewborn (2005) suggested that high-quality open-ended problems "should (1) involve significant mathematics; (2) have the potential to elicit a range of responses, from incorrect to simplistic to generalized; and (3) strike the delicate balance between providing too much information, which makes the problem restrictive and closed, and too little information, which makes the problem ambiguous" (p. 414).

The different characteristics of open-ended problems, as described in this section, are not mutually exclusive. The open-ended mathematics problems that are included in this chapter have certain general structures that emphasis various components of the problem-solving process. In summary, an open-ended problem is one that is presented in such a way that there are many possible approaches to solve it or there are many possible solutions. It is also more encompassing than typical closed problems used in many mathematics classrooms.

3 Process of Solving Open-Ended Problems

In the traditional approach, there is an inclination for students to believe that mathematics involves merely practicing one-step, two-step or many-step procedures to find answers to routine problems. However, when used regularly, open-ended problems can instill in students the idea that understanding and explanation are equally important aspects of mathematics.

While a closed problem usually has one correct solution — for example: The marks scored by 8 pupils in a mathematics test are as follows: 42, 52, 48, 44, 54, 55, 42 and 63. Calculate the mean score. — an open-ended problem is one where there are multiple correct answers and students can answer at a level that is suitable to, and represents, their current level of understanding. An open-ended problem involving the same content is shown in Figure 1. Such a problem allows students to give a range of correct solutions such as 45, 55, 42, 58, 40, 60, 70 and 30 as well as 49, 51, 49, 51, 49, 51, 49 and 51. For students who have some understanding of fractions there is opportunity to include these in their set of eight numbers as well. On the whole, student solutions also provide teachers with some insights into the student's level of understanding. This may not happen with closed problem. Thus, for mathematics teachers, the use of open-ended problem not only provides ample teaching and learning opportunities but also significant assessment information.

> List eight numbers that have a mean of 50.
> List a different set of eight numbers that also have a mean of 50.

Figure 1. An example of an open-ended problem

In the above example, the variety of solutions that students suggest allow them to contribute at their level of understanding without being considered mediocre or lacking since their solutions are correct ones. This potential of open-ended problems must not be overlooked given that many students have some form of unproductive beliefs about mathematics problems In particular, many of them believe that there is a right or wrong way to solve mathematics problems. And for many students, the latter situation is the one that they experience more often. One of the benefits of open-ended problem is that they challenge the students' belief that there is only one right technique for solving problems and this technique should be given by the teacher. Through the use of open-ended problems, teachers can help students shift their beliefs about problem solving and mathematics. The use of open-ended problems allow more discussion among students and help them recognize that, like other school subjects, mathematics is not limited to always having only one answer. By using the variety of solutions that students generate to open-ended problems as a catalyst for discussion at either whole-class or small-group levels, students are able to discuss not only their solutions but also how they arrive at their solutions. In addition, students are able to discuss other ways of finding solutions. Students are also given an opportunity to evaluate means of arriving at solutions which are more effective or efficient. This process allows teachers greater access to students' knowledge and understanding that would not otherwise be possible.

If the area of a parallelogram is 90 square metres, find its base and height.

Figure 2. Another example of an open-ended problem

The isolating of teaching and assessment is a common practice that places a lot of pressure on teachers. In contrast, other approaches to teaching and learning suggest that assessment is integral to teaching. Open-ended problems can form the basis of a lesson whereby the teacher can assess students' responses. In an open-ended problem such as the one shown in Figure 2, it is possible that within a mathematics lesson some students may give answers where the parallelogram of base 15 m and height 6 m. Thus, it shows evidence of the students' understanding of area, shape, multiplication and so forth. However, within the same class, some students may be working at a different level and have been exploring areas of parallelogram and offer answers that support their understanding of this aspect of area and shapes. Such responses may not have appeared when using closed problems. However, by posing the problem in this way, teachers are able to access more knowledge of their students' levels of understanding than would have otherwise been possible. As such, open-ended problem presents a learning situation for students and can serve as an assessment tool which gathers information on what a student has learnt and can achieve.

The use of open-ended problems allows students to solve realistic problems with incomplete information where they are required to make some assumptions about the missing information. This will provide the teacher with meaningful information on how the students manage the problem-solving process (van den Heuvel-Panhuizen, 1996). Although open-ended problems have their value in the learning mathematics for the lower secondary levels, we should not advocate using them merely because they are popular. Instead, teachers need to establish thoughtful rationale for deciding how and when to use open-ended problems in their classrooms. Schools should strongly encourage the use of open-ended problems in all aspects of mathematical instruction including the development of mathematical concepts and the acquisition of computational skills. Moreover, from the review above, it appears that open-ended problem is an effective assessment tool for enhancement of mathematical concepts, solving real-life problems and improving problem-solving ability.

4 Sample Open-Ended Problems for Lower Secondary Students

The conceptualization of the revised Singapore mathematics curriculum (Ministry of Education, 2006) is based on a framework where active learning via mathematical problem solving is the main focus of teaching and learning of mathematics. One of the main emphases of the secondary-level mathematics curriculum has been the acquisition and application of mathematical concepts and skills. While the revised curriculum continues to emphasize this, there is now an even greater focus on the development of students' ability to conjecture, discover, reason and communicate mathematics through the use of open-ended problems. The appropriate use of open-ended problem in the classrooms is a key factor in achieving the aims of the curriculum.

Since open-ended problem is an effective assessment tool for enhancement of mathematical concepts, solving real-life problems and improving problem-solving ability, it will be useful to provide four such appropriate open-ended problems for lower secondary mathematics teachers to integrate in their mathematics lessons. The following section describes four open-ended problems that can be integrated in the teaching and learning of mathematics at the lower secondary level.

Problem 1

Finding Ratio (Secondary 1)

Fill in the blanks with numbers so that the story makes sense.

The school planned a mathematics trail in the zoo. The number of buses needed to bring the pupils to the zoo is __A__. The ratio of the number of pupils to the number of teachers on the mathematics trail is __B__ to 1. There are __C__ pupils and __D__ teachers on the mathematics trail. Counting the bus drivers and principal, a total of __E__ people went on the mathematics trail.

In Problem 1, the concept of ratio is reinforced. Students who lack conceptual knowledge of ratio may make an attempt to fill in the blanks by guessing and checking. Such students may find this process of filling in the blanks tedious and cumbersome. Problem 1 requires students to make an initial decision on the number of buses required for this trip. There are many possible values for A, B, C, D and E. However, the ratio of C to D must be the same as the ratio B to 1. E must be equal to A + C + D + 1. One possible solution is A = 2, B = 10, C = 60, D = 6 and E = 69. The ratio of pupils to teachers can then be computed as 60 : 6 = 10 : 1. Another possible solution could be A = 3, B = 12, C = 96, D = 8 and E = 108. The number of buses needed and the ratio of the number of pupils to the number of teachers appear in the mathematical structure of the problem as variables. The relationship between the two makes it possible for students to relate the total number of people going on the mathematics trail. This problem can enhance students' understanding of the concept of ratio better than standard textbook problems that are typically closed, for example problems that require students to find the ratio of two or more quantities.

Problem 1 is an open-ended problem as it has many possible answers. Although students are able to form and simplify ratios, they have to make realistic assumptions and decisions (such as there is one driver per bus) in order to find the answers in this problem. Furthermore, when faced with such word problems, students should somehow represent its structure by identifying the quantities and the relationships between them in order to make a decision and to justify the decision.

Problem 2

Properties of a Rhombus (Secondary 1)

Write down as many properties and geometrical terms that you know about a rhombus.

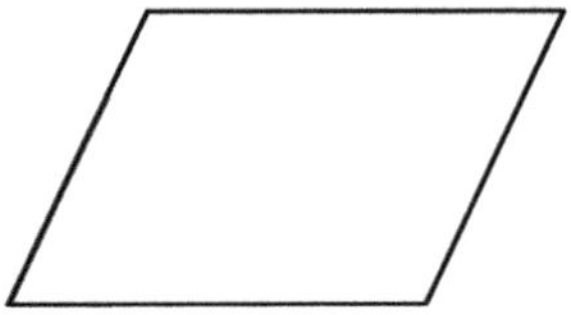

Sometimes teachers unduly pressurise students to remember properties and terminologies in geometry. However, students may not remember them. Students may also mix the properties and terminologies up easily. This is so because students have no conceptual understanding of why these properties work and the properties are not meaningful if they are merely committed to memory. Students should be strongly encouraged to use their understanding of the properties and terms to describe a geometrical situation. Therefore, the aim of Problem 2 is to enhance the students' mathematical communication where students need to express geometrical ideas of a rhombus precisely, concisely and logically. It helps students develop their own understanding of rhombus and sharpen their geometrical thinking.

Figure 3 shows a good, but not the best, response to Problem 2. The student was able to describe the main properties of rhombus and its measurement characteristics but he missed the fact that diagonals bisect each other at right angles in a rhombus out.

> Solution
>
> These are the properties and geometrical terms
>
> 1) Its opposite angles are the same.
> 2) It has 2 pairs of parallel lines.
> 3) The sum of its angles adds up to 360°.
> 4) All the lines are of the same length.
> 5) It can be formed by 2 congruent triangles.

Figure 3. A student's solution to Problem 2

Problem 3

Factorising Quadratic Expression (Secondary 2)

Is this quadratic expression $x^2 - 3x + 10$ factorisable? Explain the reasoning for your answer.

In a traditional classroom, lower secondary students have little opportunity to explain and justify the mathematical processes involved in their mathematical solutions. Sometimes they may not be able to understand what explaining their thinking meant. Although they may be able to perform certain computations, they do not know how to explain why they do them or why the procedures work. Even when a teacher insists that the students explain and justify their solution method, they may simply mimic what the teacher has said in class.

In Problem 3, students may just simply indicate that the quadratic expression is not factorisable over integers and it cannot be factorised using the so called 'cross-multiplication' method. Performing the procedure to factorise a quadratic expression in Problem 3 is easy and accessible to the vast majority of lower secondary students. In explaining their reasoning, students need to consider the constant term, 10, could be some combinations and factors of ± 2 and ± 5. They need to work through the various operations of these two numbers to be equal to coefficient x. Even though the sum of 2 and (-5) is equal -3 which is the coefficient of x, the students have to justify that the factorisable form $(x + 2)(x - 5)$ is not equal to the original quadratic expression $x^2 - 3x + 10$. Hence $x^2 - 3x + 10$ cannot be factorised over integers. This whole process of reasoning involves the flexibility in thinking about numbers that emerges with the ability to relate with the coefficients. This also creates an opportunity for students to explore and appreciate quadratic expressions.

Figure 4 shows a response that can be a good platform for the teacher to engage students in extending their thinking and reasoning as there are several ways of justifying it. However, in this problem, students should not be using the phrase 'quadratic equation' loosely. This solution shows some understanding but the generalisation is not explained adequately.

Solution

The quadratic equation $x^2 - 3x + 10$ is not factorisable.

By using the 'cross-multiplication' method :

x	-5	-5x
x	-2	-2x
x^2	+10	-3x

It can be seen that the equation cannot be factorised.

One of the reasons for not able to factorise is the signs.

In this case

x	5	5x		x	-5	-5x
x	2	2x	OR	x	-2	-2x
x^2	10	-3x		x^2	10	-3x

- The only way of a positive 10 is either a +5 and +2 or a -5 and -2.
- However, in doing so, the coefficient of the x would not be -3
- For +5 and +2, the coefficient of x will be 7 and for -5 and -2, the coefficient of x will be -7.
- Using the 'rule' of 'cross-multiplication', a quadratic equation can only be factorise if both the numerical and the coefficient of x is satisfied.

Figure 4. A student's response to Problem 3

Problem 4

Solving Simultaneous Linear Equations (Secondary 2)

John said that even though both substitution and elimination methods are used to solve simultaneous linear equations, the substitution method is a better choice for solving simultaneous linear equations when one of the variable has a coefficient 1 or -1. Do you think John is right? Give reasons to support your answers. [You may use examples to explain your answers]

Problem 4 requires students to understand that their goal was to provide enough details about both their thinking and the mathematical processes they used. This will help the teacher and other students to follow their reasoning. Problem 4 is an open-ended problem which the students are expected to apply non-algorithm thinking, to access relevant knowledge on linear equations, to apply algebraic concepts, and apply algebraic manipulation skills to simplify or change subject of a formula. They might make use examples to justify their choice. The problem assumes that students have prerequisite knowledge of solving simultaneous linear equations using substitution and elimination methods. Teachers may note that students could feel insecure about handling a mathematical problem which requires them to create another problem. Some students may have problem verbalising their methods. Teachers should strongly encourage the students to justify their decisions regardless of the decisions that they make.

Figure 5 and Figure 6 show different responses by Secondary 2 students to Problem 4. Both students show good reasoning using examples to illustrate their arguments. It demonstrates the students' complete understanding.

No, it is not very true to say that if one of the variable has a coefficient of 1 and –1, it is better to use substitution because it depends on the second equation.

$x + 2y = 30$ -----------(1)
$x + 3y = 40$ -----------(2)
} In such a case, using elimination would be much better even though the coefficient is 1.

(2) – (1) $y = 10$

$x + 2(10) = 30$

$x = 10, y = 10$

On the other hand, it could also be easier when one of the variables had a coefficient of 1 or -1 when the other equation is much more complex.

$x + 3y = 30$ ----------(1)

$5x + 10y = 250$-------(2) ⟵ This much more complex and would be better to use substitution method.

$x = 30 - 3y$

$5(30 - 3y) + 10y = 250$

$150 - 15y + 10y = 250$

$-5y = 100$

$y = -20,$

$x = 30 - 3(-20)$

$x = 90, y = -20$

Figure 5. One student's solution to Problem 4

It is right. Substitution requires the equation being substituted to be in the form of x = …….. Or y = …….., having a variable with a coefficient of ±1.
If a linear equation already has a variable with a coefficient of ±1, less is needed to be done as one would only need to move the terms instead of multiplying or dividing the whole equation so that one of the variables has a coefficient of one. The elimination method is best used when the two equations have variables with the same coefficient.

Example of use of substitution:	Example for use of elimination:
(x) − 2y = 1	6y + 5x = 16
3x + y = 17	6y − 3x = 0

Figure 6. Another students' solution to Problem 4

These four open-ended problems exemplify how open-ended problems help lower secondary students explore various types of mathematical tasks. The different open-ended problems that were discussed highlighted the different learning experiences that students gain when they work on diverse open-ended problems. This is only possible when the mathematical open-ended problems that teachers use in their classrooms go beyond computations and rote algorithms. The four problems are just first steps towards making the use of open-ended problems in the classroom a meaningful one where emphasis is on the process (reasoning and thinking) rather than the product (final answer). In addition, the consistent use of open-ended problems provides opportunities and possibilities for students to enhance their mathematical learning.

5 Conclusion

The four open-ended problems have shown that they provide mathematics teachers with quick checks into students' thinking and conceptual understanding (Caroll, 1999). They are no more time-consuming to correct than the homework exercises that teachers usually give. When used regularly, students can develop the skills of reasoning and communication in words and diagrams. Students, in presenting their solutions to others, could compare and examine each other's methods. Discoveries from such comparison and examination could allow students to modify and to further develop their own ideas in innovative ways. It is an approach that enables lower secondary mathematics teachers to teach mathematics that aligns with the spirit and intent of 'Teach Less, Learn More' (TLLM). The focus of TLLM is on thinking, reasoning and engaged learning that characterise the shift from practicing isolated skills towards developing rich network of conceptual understanding.

In addition, to assist students accustom themselves with the use of open-ended problems, assessment needs to be modified at the school level in order to focus on this new development. However, there is a need to strike a balance between basic numeracy skills, conceptual understanding and problem solving. The changes that school leaders, curriculum specialists, teachers and students need to manage for successful integration of open-ended problems into the lower secondary mathematics curriculum clearly bring a number of challenges along with them. Ultimately, the decision to use open-ended problems in the mathematics lessons is up to the teacher. It is hoped that teachers will bear in mind the appropriate use of open-ended problems by relating it to their pedagogical goals and their students' abilities.

References

Becker, J. P. & Shimada, S. (1997). *The open-ended approach: A new proposal for teaching mathematics*. Reston, VA: National Council of Teachers of Mathematics.

Chow, W. K. & Ng, Y. C. E. (Eds.), (2007). *Discovering mathematics 1A*. Singapore: Star Publishing.

Chow, W. K. & Ng, Y. C. E. (Eds.), (2008). *Discovering mathematics 2A*. Singapore: Star Publishing.

Collin, A., Brown, J. S. & Newman, S. E. (1989). Cognitive apprenticeship: Teaching the crafts of readings, writing and mathematics. In L. B. Resnick (Ed.), *Knowing learning and instruction: Essays in honor of Robert Glaser* (pp. 453-494). Hillsdale, NJ: Erlbaum.

Carroll, W. M. (1999). Using short questions to develop and assess reasoning. In L.V. Stiff & R. Curcio (Eds.), *Developing mathematical reasoning in grades K-12, 1999 Yearbook* (pp. 247-255). Reston, Va.: NCTM.

Eggen, P. D. & Kauchak, D. P. (2001). *Educational psychology: Windows on classrooms*. Upper Saddle River, N.J: Merrill Prentice Hall.

Fan, L. H. & Zhu, Y. (2000). Problem solving in Singaporean secondary school mathematics textbooks. *The Mathematics Educator, 5*(1), 117-141.

Foong, P. Y. (2002). The role of problems to enhance pedagogical practices in the Singapore mathematics classroom. *The Mathematics Educator, 6*(2), 15-31.

Hancock, C. L. (1995). Enhancing mathematics learning with open-ended questions. *The Mathematics Teacher, 88*(6), 496-499.

Kulm, G. (1994). *Mathematics assessment: What works in the classroom?* San Francisco: Jossey-Bass Publisher.

Lee, P. Y. & Fan, L (Eds.), (2007). *New express mathematics 1.* Singapore: Multimedia Communications.

Lee, P. Y. & Fan, L (Eds.), (2008). *New express mathematics 2.* Singapore: Multimedia Communications.

Leatham, K. R., Lawrence, K. G. & Mewborn, D. S. (2005). Getting started with open-ended assessment. *Teaching Children Mathematics, 11*, 413-419.

Mason, J., Burton, L. & Stacey, K. (1982). *Thinking mathematically.* London: Addison-Wesley.

Ministry of Education. (1990). *Mathematics syllabus (Lower Secondary).* Singapore: Curriculum Planning Division.

Ministry of Education. (2000). *Mathematics syllabus (Lower Secondary).* Singapore: Curriculum Planning Division.

Ministry of Education. (2006). *Secondary mathematics syllabus.* Singapore: Curriculum Planning and Development Division.

National Council of Teachers of Mathematics (2000). *Principles and standards for school mathematics.* Reston, VA: Author.

NSW Board of Studies (2002). *Mathematics K-6 syllabus 2002.* Sydney, New South Wales, Australia: Author.

Sin, K. M. & Chip, W. L (Eds.), (2007). *Mathematics matters 1.* Singapore: Panpac Education.

Sin, K. M. & Chip, W. L (Eds.), (2008). *Mathematics matters 1.* Singapore: Panpac Education.

Sullivan, P. & Lilburn, P. (2005). *Open-ended maths activities: Using 'good' questions to enhance learning.* Melbourne: Oxford University Press.

van den Heuvel-Panhuizen, M. (1996). *Assessment and realistic mathematics education.* Utrecht: CD-B Press/Freudenthal Institute, Utrecht Univerisity.

Chapter 13

Arousing Students' Curiosity and Mathematical Problem Solving

TOH Tin Lam

Problem solving is the heart of the Singapore school mathematics curriculum. While teaching students mathematical problem solving, which includes problem solving heuristics and thinking skills, it is important that teachers arouse their curiosity and engage them. They may do this by introducing mathematics through daily life activities, modifying their normal approach of classroom teaching, making classroom mathematics relevant in real-life and elaborating the less prominent mathematical results. This chapter illustrates through examples how each of the above may be achieved in the secondary mathematics classroom.

1 Introduction

Since the publication of Polya's first book about solving mathematics problems (Polya, 1945) there has been much interest in mathematical problem solving. From the 1980s, there has also been a world-wide push for problem solving to be the central focus of school mathematics curriculum. For example, in the United States, the National Council of Teachers of Mathematics (NCTM) in their document on the principles and standards for school mathematics stated that: "[p]roblem solving should be the central focus of the mathematics curriculum" (NCTM, 2000, p.52).

In line with global trends in mathematics education, mathematical problem solving was established as the primary focus of the Singapore school mathematics curriculum since the 1980s. The primary aim of the curriculum is to develop students' ability to solve mathematics problems. The framework of the Singapore mathematics curriculum is shown in Figure 1.

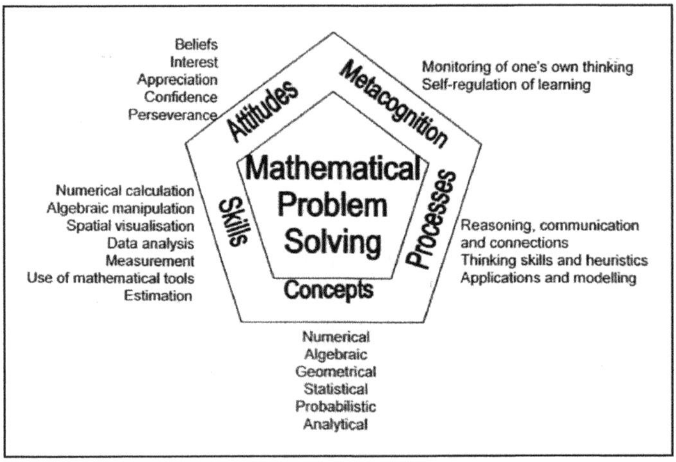

Figure 1. Framework of the Singapore school mathematics curriculum

It is apparent from Figure 1, that the Singapore mathematics school curriculum framework highlights the dependence of development of mathematical problem solving ability on five inter-related components: Concepts, Skills, Processes, Attitudes and Metacognition (Ministry of Education, 2006). Teachers often emphasize Concepts, Skills and Processes for successful mathematical problem solving. They often fail to note that Metacogniton and Attitudes are equally important for engagement in problem solving. Only recently, increasing attention has been placed on the Metacognitive aspect of mathematical problem solving (see, for example, Toh, Quek and Tay, 2008a; 2008b).

Other than being concerned about students' ability to solve problems, how often do teachers think about the following related to their classroom practice:

- How do my students feel about solving mathematics problems (for example, do they feel unduly stressed when solving non-routine mathematics problems)?

- Are my students really interested in solving the mathematics problems (or are they only interested in preparing for the national examinations)?
- Do my high achieving mathematics students really enjoy solving the challenging mathematics problems I assign them?
- Do my students really appreciate the mathematics they learn?

The above questions relate to the domain of the affective aspects of mathematics learning. i.e. the Attitude component of mathematical problem solving. Generally, the Attitude component is given relatively less attention compared to the other four components.

2 Arousing Students' Curiosity in Mathematics and Problem Solving

Any review of literature on mathematical problem solving would inevitably start with Polya's conception of solving mathematics problems (Polya, 1945). Polya did not use the term "problem solving", but discussed "studying the methods of solving problems". According to Polya, solving a problem would mean "finding a way out of a difficulty, a way around an obstacle, attaining an aim which was not immediately attainable" (Polya, 1981, p.ix). Polya's model of solving problems, which forms the foundation of the Singapore mathematics curriculum, can be presented as consisting of four main stages: (1) Understanding the Problem; (2) Devising a Plan; (3) Carrying out the Plan; and (4) Checking and Extension.

What, then, is a mathematics problem? According to the definition by Lester (1978), which is generally accepted by mathematics educators, a problem is a situation in which an individual or group is called upon to perform a task for which there is no readily accessible algorithm which determines completely the method of solution. Lester (1980) adds that this definition assumes a desire on the part of the individual or group to perform the task.

As a corollary to Lester's definition of a *mathematical problem*, what is considered a mathematical problem to one student might NOT be a *problem* to another student; if (i) the latter has a ready algorithm to solve the questions but not the former; and (ii) the former is interested in solving the problem but not the latter. Thus, if Lester's definition of a *mathematics problem* is adopted, in teaching mathematical problem solving, teachers must be able to draw on a rich source of questions and contexts that are exciting and "new" to their students, and to develop interest in their students in solving the questions.

Educators worldwide have been discussing ways to develop students' interest in learning mathematics and mathematical problem solving. For example, the Chinese teachers have been talking about developing students' interest in mathematics by relating the content to real-life situations (Correa, Perry, Sims, Miller, Kevin and Fang, 2008). Research has also shown that teachers' use of stimulating teaching methods would go a long way to sustain and motivate students' interest in learning mathematics (Akinsola, Animasahun; 2007). In particular, using hands-on activities in real-life situations could help the students see the relevance and feel the power of mathematics (Morita, 1999; Mitchell, 1994). The above are some examples from a long list of studies and researches on the different approaches to develop students' interest in mathematics learning.

Underlying all the above discussion of ways to develop students' interest in mathematics is one of the key psychological aspects – arousing the students' curiosity in mathematics. What, then, is curiosity? Curiosity is not the mere wonder of a feat or an event. According to Schmitt and Lahroodi (2008), curiosity is a motivationally original desire to know. This desire arises and, in turn, sustains one's attention and interest to know. Curiosity is a characteristic that is often observed in our students.

The importance of curiosity cannot be overemphasized. Curiosity can lead students to explore new ideas in mathematics (Gough, 2007). Some researchers have even asserted the importance of curiosity as an important link to an individual's lifelong learning (Fulcher, 2008). However, curiosity has not received much academic interest until recent years.

In this chapter, we are going to discuss the different ways of arousing students' curiosity through daily activities and events which are linked to mathematical concepts and principles. Broadly speaking, as seen by the author, these can be classified into the following categories:

- Introducing mathematics through daily activities;
- Modifying the normal approach of classroom teaching;
- Making classroom mathematics relevant in real-life; and
- Elaborating on the less prominent results.

2.1 *Introducing mathematics through daily activities*

There are a lot of opportunities to explore "mathematics" in our daily activities. Teachers can employ all these opportunities to arouse their students' curiosity in the subject. In this section, we propose some activities related to two things students encounter in their daily lives: (1) calendar, (2) page numbers of books.

2.1.1 *Calendar*

There is a good deal of mathematics in the calendar. However, students might not have linked the many day-to-day events associated with the calendar to school mathematics. A search of the available websites show numerous sites which offer interesting mathematical tasks based on the calendar. In this section, we shall list several mathematical tasks that are related to the calendar.

As an example of the mathematics of calendar, the calculation of the days of the same date for each month is an arithmetic problem related to the remainder of an integer by the number 7. For example, if it is known that 1st January 2008 falls on Tuesday, the mathematics classroom teacher can challenge the students to find the day on which the 1st of every subsequent month falls *without* telling them the related mathematics. The teacher could get students to think of how to solve the problem, and relate this to the mathematics (in particular, arithmetic) that they have learnt in their mathematics classrooms.

Teachers can also generate higher order thinking questions using the calendar in many ways after their students have understood the basic mathematical calculations. For example, the following question can be used as an investigative task for the students, leading them to experience and discover the underlying mathematical principle associated with arithmetic:

> Many people believe that Friday 13th is very inauspicious. Is it possible for you to find a year in which there is no such "inauspicious" day (that is, 13th of every month falls on any day other than Friday)? What is the maximum number of such "inauspicious" days in any particular year? Can you tell me your answers without referring to all the past and future calendars?

There are many other potentially interesting problems that can be related to the calendar. For example, Bastow, Hughes, Kissane and Randall (1986, p.11, n.19) demonstrated some interesting activities for students for mathematical investigation:

> Someone said: "We can use this year's calendar again in a few years from now." Investigate. (Bastow, et. al., p.11)

Generalizing the above problem, a teacher could get the students to think:

> If you are a manufacturer of calendars, must you produce a "new" calendar every year? Or do you observe that after a certain number of years, the same calendars can be used again?

Even a task like getting students to figure out the day of the week which they were born could be a motivating one to arouse the students' curiosity in the mathematics related to daily life:

> [Without the use of calendars, o]n what day of the week were you born? (Bastow, et. al., p.19).

How can a teacher implement some or part of the above activities on calendars in a classroom setting? As an illustration, the teacher could use these activities as motivation to start a new chapter in class, especially topics related to arithmetic or counting, to arouse the students' curiosity of the application of mathematics in their daily lives. The teacher could also use these activities as part of the enrichment classes supplementing the usual classroom lessons.

Extracts of the above activities could also be used for out-of-class activities: for example, teachers could either incorporate some of the abovementioned activities in inter-disciplinary activities, for example, through an activity that incorporates National Education, during which students are introduced to some days of special significance to the nation, or during activities in which students are induced more into the culture of the school and the early history of the founding of their secondary schools.

2.1.2 *Page numbers of books*

In Singapore secondary schools (Years 7 to 10), students use books (textbooks, exercise books or notebooks) for most of their lessons. However, not many students would have noticed that there is a lot of interesting mathematics related to page numbers of the books. In this section, we shall demonstrate with an illustration the mathematics associated with the page numbers of a book. Consider the case of an A4-sized book, in which the pages are printed on A3 papers. In this way, one piece of A3 paper of the book consists of four printed pages of the book.

An example of a suitable mathematics problem is finding the page numbers of all the four pages that are printed on the same piece of paper. The students could be guided through a series of investigative activities to discover that the sum of the page numbers of every four pages printed on the same sheet of paper is always constant. If a book has $4n$ pages (the number of pages in this context is always a multiple of 4), then the sum of every four pages on the same sheet of paper is always $8n + 2$.

Other potentially interesting problems related to the page numbers of a book are given below. However, it is entirely up to the professional judgment of the classroom teachers to generate as many other interesting and creative mathematics problems related to the page numbering of a book.

1. A boy opens up the middle pages of two books and finds that the product of the two numbers of the middle pages is 42. How many pages does the books have?
2. A book has n pages. The book is numbered from page 1 to page n. Mary added up all the page numbers and got the sum equal to 3250. However, she added up the numbers wrongly because there is a particular page number that she added up twice. What is the page number that she counted twice in her calculation?
(Chua, Hang, Tay and Teo, 2007, p.120)

While the above two examples are taken from the Singapore Mathematical Olympiads, they can be used equally well in the usual classroom setting as a mathematical investigation task to stretch students' mathematical thinking.

2.2 *Modifying the usual approach of classroom teaching*

Research has shown that mathematical problem solving and acquisition of mathematics concepts are difficult for students. The use of appropriate instructional strategies is crucial to the students' understanding of mathematical concepts (Akinsola, 1994, 1997). For effective instruction to take place, teachers are required to step outside the realm of their own personal experience into the world of their students (Brown, 1997). As such, it is not uncommon that teachers modify the usual way of classroom lesson delivery to meet the learning needs of their students. We illustrate how this may be done with a few examples taken from the Singapore secondary school mathematics curriculum (Ministry of Education, 2006).

2.2.1 *An example from teaching mensuration*

Consider an example of a mathematics teacher attempting to teach his/her students the formula for the circumference of a circle as 2π multiplied by the radius of a circle. Let us consider two approaches of teaching: (1) the teacher may get the students to memorize the formula and then apply the formula to solve the textbook exercises, or (2) the teacher could get the students to be involved in simple tasks of measuring the circumference and diameter (or radius) of many circles of various sizes and check that the ratio of the circumference to the diameter (or radius) always has a constant value of π (2π, respectively). While the message of the formula for the circumference of the circle might be conveyed to students in both cases, the effect might be different. In the second case, the teacher is likely to have aroused the students' curiosity in the mathematical concept of the formula regarding the circumference of the circle.

2.2.2 *Another example from teaching algebraic manipulation*

We consider another example on algebraic manipulation. Students generally find the learning of algebra difficult. Students find some algebraic rules generally more difficult to learn compared with the other rules (Kirsher and Awtry, 2004). Visually salient rules, due to their visual coherence that makes them seem more "natural" and believable, are easier for students to learn. On the other hands, algebraic rules that are less visually salient are more difficult for students to learn. An example of such a rule is the expansion

$$(a + b)^2 = a^2 + b^2 + 2ab.$$

There are several ways to teach these less visually salient rules: (1) the teacher could get the students to memorize the formulae and practise sufficiently many exercises on algebraic expansion and factorization applying these rules; (2) the teacher could explain why such rules work. For example, Yeap (2007) provides worksheets to demonstrate to students that $(a+b)^2$ is not equal to a^2+b^2 through involving students trying out numerical examples (Yeap, 2007; p.37) or through geometrical interpretation (Yeap, 2007; p.41 – 42).

Here, the author proposes another way to arouse students' interest in learning these algebraic rules, which is through "impressing" the students with these algebraic rules by guiding them to experience the "power" of these algebraic rules. For instance, a teacher can challenge his or her students on the speed of performing the following computations:

a. $(2007)^2 - 2 \times 2007 \times 2006 + (2006)^2$
b. $(2007)^2 - (2006)^2$
c. $(45)^2 + 2 \times 45 \times 55 + (55)^2$
d. $(45)^3 + 3(45)^2(55) + 3(45)(55)^2 + (55)^3$
e. $(0.5)^3 - 3(0.5)^2(0.4) + 3(0.5)(0.4)^2 - (0.4)^3$

Simply by using appropriate algebraic identities without the use of calculators, the correct values of the above expressions can be found almost instantly; On the other hand, evaluating the above expression mechanically without algebraic rules (or even the use of calculators) will be far less efficient. We could use the above type of examples in the usual mathematics classrooms to illustrate to students that, under certain circumstances, use of algebraic identities could be a more efficient tool in computation using a calculator.

2.2.3 *Extension from pattern gazing*

Pattern gazing, or more commonly known as *observing number patterns* in the Singapore mathematics curriculum, is taught in the Singapore mathematics classroom at the lower secondary levels (Years 7 and 8). Through pattern gazing, students are exposed to some problem solving heuristics, such as forming conjecture, generalizing. However, teachers may decide not to stop here; they could use these pattern gazing activities to introduce to their students many interesting aspects of the subject, thereby arousing their curiosity to find out more such activities. One example of an activity used by the author during an enrichment course with a group of lower secondary students is appended below.

Fill in the blanks below:

1 + 3 = ____

1 + 3 + 5 = ____

1 + 3 + 5 + 7 = ______

1 + 3 + 5 + 7 + 9 = _______

What is the sum of the first 100 odd positive integers?

____

1 + 3 + 5 + 7 + 9 +......... + 99 = _____

Figure 2. One activity on pattern gazing

Most students would have realized, after attempting to complete the first two or three lines of Figure 3, that the sum of the first *n* odd positive integers is somehow a perfect square. The higher achieving students would be able to conclude, through completing the first four lines in Figure 3, that *the sum of the first 100 odd positive integers is 10000* and, after some calculations, would be able to conclude that the value of the sum of odd integers 1 + 3 + 5 + 7 + 9 + ... + 99 is 2500.

One question that the teachers could get their students to ponder on: Is the fact that "the sum of the first *n* odd positive integers a perfect square" a mere "coincidence", or is there any rigorous explanation besides pattern gazing?

Such an activity would be redundant if we subscribe to the belief that people are "minimalist information processors" who are unwilling and uninterested to devote much effort to process this type of arguments (Stiff, 1994). However, if teachers believe that students could be challenged to explore into such unknown territories, they could stretch their students to find a plausible pictorial explanation for the above fact. For example, Nelson (1993) gave two plausible "proofs without words" in his book (Nelson, 1993, pp. 71 – 72). One of such possible explanation is shown in Figure 3 below.

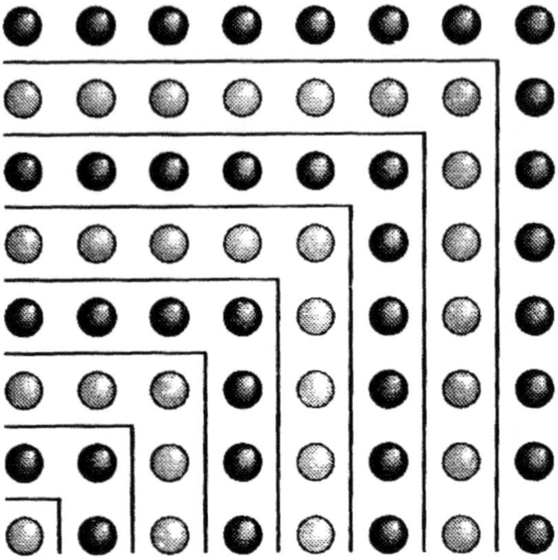

$$1+3+5+\ldots+(2n-1)=n^2$$

Figure 3. One explanation for the formula of the sum of odd integers in Nelson (1993)

The above sample activity illustrates that the usual activity used as pattern gazing in the teaching of the usual mathematics curriculum can be *extended* (beyond just getting the correct answers) to arouse students' curiosity in the subject in general and, in particular, mathematical problem solving.

Many other number patterns, for example, the sum of positive integers, the sum of squares of positive integers and the sum of cubes of positive integers, can be used for pattern gazing and also be used to involve the students in such arguments and curiosity into mathematics (Toh, 2007).

2.3 *Making classroom mathematics relevant in real-life*

Researchers and educators agree on the importance of relating mathematics to real life applications (see, for example, Albert and Antos,

2000). In fact, educators are prepared with ways to answer students' questions about when they will use the mathematics they learn in the classrooms (Gough, 1998). In this section, we shall illustrate with examples how some of the mathematics our students learn in secondary school (Years 7 to 10) relates to the real world, and may serve as motivation.

2.3.1 *Two examples from mensuration: Area of trapezium and volume of frustum*

The formula for the area of a trapezium is well-known. In Lee (2007, p.108), teachers were expected to be able to derive the formula of the area of a trapezium. The derivation of the formula could be done by placing two congruent trapeziums as shown in Figure 4.

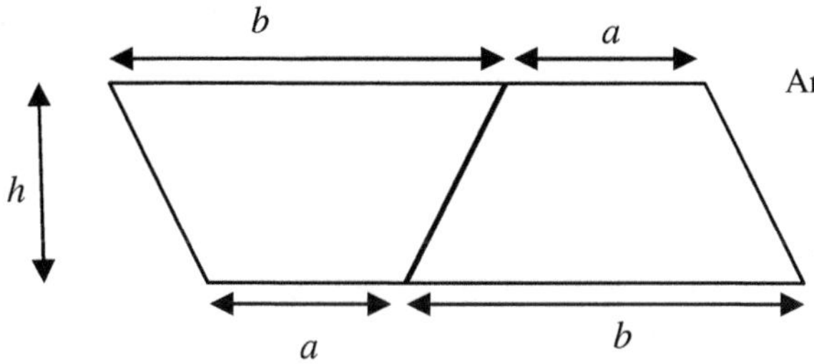

$$\text{Area} = \frac{1}{2}h(a+b)$$

Figure 4. Two congruent trapeziums placed together to obtain a parallelogram

One might ask: are there other situations in the real world that the method used to find the formula for the area of a trapezium (by using two congruent objects to form a familiar object whose formula is known) can be applied to? Teachers can excite their students to explore using the method in other new situations. As an illustration on how the method of deriving the formula for the area of a trapezium can be extended, teachers can challenge their students to find the volume of the solid shown in Figure 5.

Figure 5. Photograph of a short pillar in front of a lift at Changi Airport Terminal Three

A student who has understood the principle underlying the method of finding the area of a trapezium would be able to extend the method to find the volume of the short pillar in Figure 5 above; he/she would observe that two congruent volumes placed in the position of Figure 6 would end up in forming a circular cylinder, and hence obtain the formula for the volume of the short pillar as being half of the volume of the entire cylinder in Figure 6.

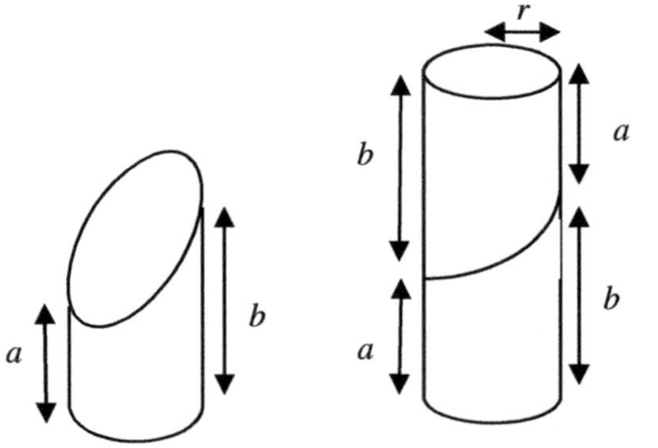

Volume of the solid =

$$\frac{1}{2}\pi r^2(a+b)$$

Figure 6. Diagram illustrating the method to find the volume of the solid in Figure 5

In the Singapore secondary school mathematics curriculum, students are required to acquire concepts related to similar figures and similar solids and their properties, areas and volumes of standard figures and solids as required by the syllabus. However, in real life situations, most figures and solids do not belong to the "standard" shapes taught in the school mathematics curriculum. Teachers could challenge their students to find areas and volumes of other "non-standard" figures and solids, based on what they have learnt in their curriculum about the areas and volumes of standard figures and solids, and concepts of similarity and congruency.

As an illustration, teachers could get their students to try finding the volume of the hanging lamp (in the shape of a frustum) shown in Figure 7 below.

Figure 7. Photograph showing the outlet of a shop with hanging lamps in the shape of a frustum

In finding the volume of a frustum, students could come to realize that they would need to apply concepts of similar triangles and formula for the volume of a circular cylinder to find the volume of a frustum.

2.3.2 *Two more examples from the use of the common logarithms*

Logarithm is taught in the additional mathematics curriculum to upper secondary students (Years 9 and 10). However, the application of logarithm to real-life situations is rarely highlighted during the curriculum. Teachers should note that one can easily find applications of logarithm to the real world situation, if one appreciates that logarithms are used to handle "unusually large or unusually small numbers". For instance, the Richter scale of measuring earthquake is an example of a logarithmic scale.

Students doing science, may recognize the use of logarithms for the measurement of the acidity of a solution, the pH scale. The pH of a solution can be calculated by

$$pH = -\lg [H^+],$$

where $[H^+]$ denotes the concentration of hydrogen ions in mol/dm^3 of the solution. Based on this formula, the teacher can challenge the students with questions related to the formula and the concept of logarithm such as the following questions.

> 1. Based on the formula for pH above, is it possible for a solution to contain <u>no</u> hydrogen ion?
> 2. When a solution is neutral (neither acidic nor alkaline), its pH value is 7. What is the concentration of the hydrogen ion?
> 3. Suppose you are given an acid. If you add a lot of water (pH 7) into it, what will the pH of the acid reach? What if you add even more water? Explore.

Such questions would lead students to greater in-depth thinking about the mathematics formula in the context of another discipline; further, such inter-disciplinary tasks might show students greater connection across the different discipline instead of perceiving the disciplines as isolated bits of knowledge.

Another illustration on the use of logarithm in the mathematics classroom is as follows. Using a scientific calculator, teachers could lead their students to explore the common logarithm of many different integers. For example,

$$\lg 2 = 0.301\ldots.$$
$$\lg 9 = 0.954\ldots$$
$$\lg 78 = 1.892\ldots$$
$$\lg 97 = 1.986\ \ldots$$
$$\lg 123 = 2.0899\ldots$$
$$\lg 987 = 2.9843\ldots$$

Students could come to discover that the common logarithm of a whole number could be related to the number of digits that number is expressed in its decimal representation. Eventually, students could even be challenged to answer the following questions related to logarithms.

Below are two sample questions which could relate the application of logarithms to solve some higher order thinking questions. These questions could be used to induce students to mathematical problem solving.

1. The number 2^{7894} is an extremely large number which cannot be displayed on a scientific calculator. However, if you are able to write down the number in its decimal representation in full, how many digits will the number have?
2. Let us assume that you have done the above (i.e. write down the number in its decimal representation in full). Can you tell me what is the left most digit?

2.4 *Elaborating on the less prominent results*

There are some mathematics formulae and results in the secondary mathematics curriculum which are not elaborated conceptually during classroom lessons. However, the students are expected to memorize these results by rote and apply them to solve typical examination questions. Teachers can make use of this opportunity to arouse students' curiosity and excite them about the mathematics (Chen and Toh, 2008). In this section, we shall illustrate two such examples.

2.4.1 *Volume and surface area of a sphere*

Many secondary school textbooks used in Singapore schools do not elaborate on the derivation of the formulae for the volume or the surface area of a sphere. For students who are curious on how these formulae were derived, teachers could refer them to the history of mathematics so that they may appreciate how these formulae were derived historically (see Dunham, 1994, p225 – 236).

Even if the teacher considers the derivation of the formulae for the volume and the surface area of a sphere difficult for the average students, the teacher can establish a relationship between the two formulae. For example, he or she can challenge the students to contemplate on why the relation

$$V = \frac{1}{3}Ar,$$

where V = volume of sphere, A = surface area of the sphere and r = radius of the sphere, is always true for all spheres. A careful consideration of the following diagram from Billstein, Libeskind and Lott (2001) in Figure 8 below indeed shows that the relation is valid.

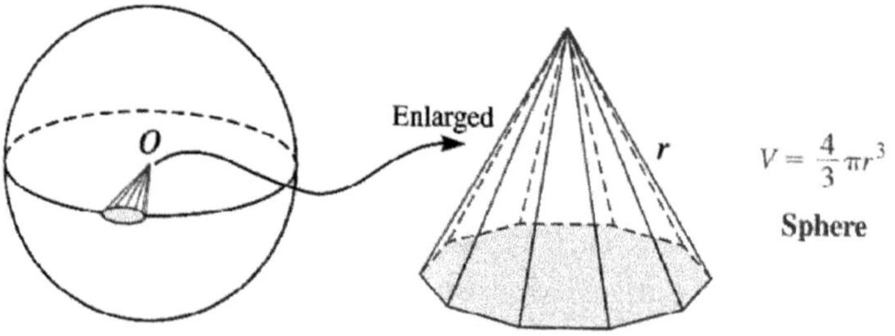

Figure 8. Diagram illustrating the relation between the volume and surface area of a sphere (Billstein, Libeskind and Lott, 2001, p.653)

The above example illustrates that teachers could use examples of formulae and results which are not usually elaborated in the usual classroom teaching and challenge their students and arouse their interest in mathematics and problem solving.

We conclude this chapter with one final example on the teaching of differentiation in the additional mathematics curriculum. While teaching the students the first principle of differentiation, teachers could consider illustrating the concept using more "lively" examples on top of the classical examples from the textbooks. They could get their students to attempt a plausible explanation of why the derivative of volume of spheres (with respect to its radius) indeed gives the surface area of the spheres; another observation is that the derivative of the area of circles with respect to their radius gives its circumference. This is not a mere coincidence; there is a plausible explanation using the first principle of derivative (see Chen and Toh, 2008). This will be a suitable type of activity to challenge students to apply what they have learnt to their encounters in other areas of mathematics, while giving a plausible explanation.

3 Conclusion

Before concluding this chapter, there are two more cautions for the classroom teachers in using the ideas presented in this chapter. Firstly, in line with the spirit of problem solving, the process of problem solving should NOT be viewed as the process of searching for algorithms or attempts to find more explicit rules or procedures to solve a (non-routine) problem. There is always a risk that teachers might use the examples illustrated in this chapter (and other examples) to push for solutions without focusing on the problem solving processes or strategies. It is thus extremely important that teachers appreciate that the importance of problem solving lies not on the final solution but more on the processes.

Secondly, some examples illustrated here might not be able to arouse students' curiosity as these problems may be some routine questions for them. Thus, it is also crucial that the teachers know their students' inclination well, are able to choose appropriate examples that are new to their students, and are able to excite their students in these selected problems.

While the literature on mathematics education abounds with research and studies on teaching students mathematical problem solving, the area on arousing students' curiosity (thereby sustaining their interest in mathematical problem solving) is relatively less explored. Taking care of the students' affective domain in mathematical problem solving could be an equally important area of mathematical problem solving. In this chapter, we have presented generally four approaches of arousing students' curiosity in mathematics and illustrated each with examples from the Singapore secondary school mathematics curriculum.

References

Akinsola, M.K. (1994). Effect of enhanced mastery learning strategy on achievement and self-concept in mathematics. *Journal of the Science Teachers Association of Nigeria, 29* (1& 2), 65-71.

Akinsola, M.K. (1997). Reward system in cooperative learning as a factor affecting mathematics achievement. *Journal Research in Education, 1* (2), 122-128.

Akinsola, M.K., & Animasahun, I.A. (2007). The effect of simulation-games environment on students achievement in and attitudes to mathematics in secondary schools. *Turkish online journal of educational technology, 6* (3).

Albert, L.R., & Antos, J. (2000). Daily journals connect mathematics to real life. *Mathematics teaching in the middle school, 5*(8), 526-531.

Bastow, B., Hughes, J., Kissane, B., & Randall, R. (1986). *Another 20 mathematical investigations.* Australia: The Mathematical Association of Western Australia.

Billstein, R., Libeskind, S., & Lott, J.W. (2000). *A problem solving approached to mathematics for elementary school teachers.* New York: Addison-Wesley & Longman.

Brown, B.L. (1997). New learning strategies for generation X. *Eric Digest* No. 184.

Chen, W., & Toh, T.L. (2008). On relationship between volume and surface area and perimeter: arousing students' curiosity in the world of mathematics. *Mathsbuzz, 10*(1), 2-3.

Chua, S.K., Hang, K.H., Tay, T.S. & Teo, T.K. (2007). *Singapore mathematical Olympiads: 1995-2004.* Singapore: Singapore Mathematical Society.

Correa, C.A., Perry, M., Sims, L.M., Miller, K.F., & Fang, G. (2008). Connected and culturally embedded beliefs: Chinese and US teachers talk about how their students best learn mathematics. *Teaching and Teacher Education: An international journal of research and studies, 24* (1), 140-153.

Dunham, W. (1994). *The mathematical universe: an alphabetical journey through the great proofs, problems, and personalities*. New York: John Wiley.

Fulcher, K.H. (2008). Curiosity: a link to assessing lifelong learning. *Assessment Update, 20*(2), 5-7.

Gough, J. (1998). Maths? Huh! – when will I ever use it? – Some reflections. *Australian Mathematics Teacher, 54*(4), 12-16.

Gough, J. (2007). How small is a billionth? *Australian Mathematics Teacher, 12*(2), 10-13.

Kirshner, D., & Awtry, T. (2004). Visual salience of algebraic transformations. *Journal for Research in Mathematics Education, 35*(4), 224-257.

Lee, P.Y. (Ed, 2007). *Teaching secondary school mathematics: a resource book.* Singapore: McGraw-Hill.

Lester, F.K. (1978). Mathematical problem solving in the elementary school: Some educational and psychological considerations. In L.L. Hatfield & D.A. Bradbard (Eds.), *Mathematical problem solving: Papers from a research workshop* (pp. 53-88). Columbus, Ohio: ERIC/SMEAC.

Lester, F.K. (1980). Research on mathematical problem. In R.J. Shumway (Ed.), *Research in Mathematics Education* (pp. 286-323). Reston, Virginia: NCTM.

Ministry of Education. (2006). *A Guide to Teaching and Learning of O-Level Mathematics 2007*. Singapore: Author.

Mitchell, M. (1994). Enhancing situational interest in the mathematics classroom. *Research Report for a meeting*. New Jersey.

Morita, J.G. (1999). Capture and recapture your students' interest in statistics. *Mathematics Teaching in the Middle School, 4*(6), 412-418.

NCTM (National Council of Teachers of Mathematics). (2000). *Principles and standards for school mathematics*. Reston, VA: NCTM.

Nelson, R.B. (1993). *Proofs without words: Exercises in visual thinking*. The Mathematical Association of America.

Polya, G. (1945). *How to solve it*. Princeton: Princeton University Press.

Polya, G. (1981). *Mathematics discovery: On understanding, learning and teaching problem solving (combined edition).* New York: John Wiley & Son.

Schmitt, F.F., & Lahroodi, R. (2008). The epistemic value of curiosity. *Educational Theory, 58*(2), 125-148.

Stiff, J.B. (1994). *Persuasive communication*. New Work: Guilford.

Toh, T.L. (2007). An in-service teachers' workshop on mathematical problem solving through activity-based learning. *Journal of Science and Mathematics Education in Southeast Asia, 30*(2), 73-89.

Toh, T.L., Quek, K.S., & Tay, E.G. (2008a). *Problem solving in the mathematics classroom (junior college)*. Singapore: National Institute of Education.

Toh, T.L., Quek, K.S., & Tay, E.G. (2008b). Mathematical problem solving - a new paradigm. In Vincent, J., Pierce, R., Dowsey, J. (Eds). *Connected Maths* (pp. 356-365). Victoria: The Mathematical Association of Victoria.

Yeap, B.H. (2007). Teaching of algebra. In Lee, P.Y. (Ed). *Teaching secondary school mathematics: a resource book* (pp. 25-50). Singapore: McGraw-Hill.

Part IV

Future Directions

Chapter 14

Moving beyond the Pedagogy of Mathematics: Foregrounding Epistemological Concerns

Manu KAPUR

The 2009 yearbook of the Association of Mathematics Educators (AME) of Singapore presents a good start to what is envisioned to be a series on mathematics education. In this chapter, I lay out the possibility space of critical issues that the yearbook could address in the coming years. In the main, I draw on the folk categories of "learning about" a discipline and "learning to be" a member of the discipline (Thomas & Brown, 2007) to propose a move beyond the *pedagogy* of mathematics to include the *epistemology* of mathematics as well. To accomplish this move, I propose a focus on three essential (but by no means exhaustive) research thrusts: a) understanding children's inventive and constructive resources, b) designing formal and informal learning environments to build upon these resources, and c) developing teacher capacity to drive and support such change.

1 Introduction

This yearbook presents an important milestone for the AME of Singapore. The yearbook's focus on mathematical problem solving is apt, and the chapters in the book represent a diverse set of emphases on pedagogy and practice. My aims for this chapter are modest and twofold. I will start by stepping back and briefly examining what it means to learn mathematics. This is important because it sets the stage for the second aim wherein I will derive implications for mathematics education research and practice, and in the process, lay out what I believe to be some (but not all) of the critical areas and issues that the AME yearbooks could focus on.

2 What does it mean to learn Mathematics?

This seemingly simple question has important implications. To learn mathematics, one must minimally be able to understand mathematical concepts, strategies, and procedures, and apply them to solve a diverse set of problems, simple or complex, routine or non-standard. Much research and practice is geared towards developing mathematical problem solving skills in children. Indeed, this AME yearbook also focuses on problem solving. Here, the concerns are largely of the form: What is the nature of children's mathematical understandings? How can we teach mathematical concepts better? What kinds of problems, activities, and tools are best suited for understanding mathematical concepts? What curricular design principles are more effective than others, and so on? Taken together, these concerns are largely pedagogical; their focus is mainly on *learning about* mathematics, which is necessary but not sufficient.

Part of learning mathematics, and arguably the more important part perhaps, is to engage in the practice of mathematics akin to that of mathematicians. This involves *learning to be* like a member of the mathematical community (Thomas & Brown, 2007). But what does mathematical practice entail? Inventing representational forms, developing domain-general and specific methods, flexibly adapting and refining or inventing new representations and methods when others do not work, critiquing, elaborating, explaining to each other, and persisting in solving problems define the epistemic repertoire of mathematical practice (diSessa & Sherin, 2000). Learning to be like a mathematician is to learn and do what mathematicians do; it involves a "mathematical" way looking at the world, understanding the constructed nature of mathematical knowledge, and persisting in participating in the construction and refinement of mathematical knowledge. These concerns clearly foreground the epistemological aspects of mathematical practice.

Therefore, from this brief examination, it follows that learning about mathematics primarily foregrounds a *pedagogical* concern whereas learning to be like a mathematician foregrounds an *epistemological* concern. Both concerns are important but the latter remains much neglected and, therefore, needs to be addressed with greater force going forward. That said, it is important to note that learning about and learning to be are inextricably dialectical; the distinction between them is merely an analytical device I employ here for the purposes of this chapter.

3 Implications for mathematics education research and practice

Because much attention has been devoted to the pedagogical concern, I will focus my attention on the epistemological concern. This naturally begs the question: *How do we design opportunities and learning experiences for students for them to understand, learn, and do (at least in some ways) what mathematicians do*? To be clear: I'm not suggesting that we need to prepare all children to become mathematicians. What I am suggesting is that if learning to be like a mathematician involves participating in the processes of inventing and refining representational forms and methods, collaborating and critiquing each other, persisting in solving problems, and a way of working with mathematical knowledge, then we need to design opportunities for student to be able to engage in these processes; processes that mirror the actual practice (diSessa & Sherin, 2000; Thomas & Brown, 2007).

3.1 *Understanding children's inventive and constructive resources*

It follows from above that if we are to engage children in the processes of invention of and "play" with representational forms and methods, we need to at least be able to at least answer some essential questions: What is the nature of children's inventive and constructive resources? What kinds of tasks, activities, and classroom cultures are more effective than others at uncovering these resources?

A growing body of research has demonstrated that children have intuitive yet sophisticated set of constructive resources to generate representations and methods to solve problems without any direct or formal instruction (e.g., diSessa, Hammer, Sherin, & Kolpakowski, 1991; Hesketh, 1997; Kapur, 2008; Slamecka & Graf, 1978; Schwartz & Martin, 2004). For example, diSessa et al. (1991) found that when sixth graders were asked to invent static representations of motion, students generated and critiqued numerous representations, and in the process, demonstrated not only design and conceptual competence but also meta-representational competence. Likewise, Schwartz and Martin (2004) demonstrated a hidden efficacy of invention activities when such activities preceded direct instruction (e.g., lectures), despite such activities failing to produce canonical conceptions and solutions during the invention phase.

Going forward, therefore, we need to design opportunities for students to leverage their constructive resources to invent, play with, and refine representational forms and methods. Such efforts will necessarily involve a variety of tools (e.g., computers, modeling and simulation tools, etc.) and activity structures (e.g., collaboration) because each has its own affordances and constraints (Greeno, Smith, & Moore, 1993). We need research that seeks to understand the interplay between the designed affordances and constraints and their influence on the learning of mathematics (Greeno et al., 1993).

3.2 *Formal and informal learning designs that help build upon children's constructive resources*

Uncovering children's constructive resources is necessary but not sufficient because their inventions are rarely the canonically correct structures (e.g., representational forms and methods). We need to design learning (environment, tasks, activity structures, etc.) so as to be able to build upon their generative structures, compare and contrast them with each other and with the canonical structures. Again, this only begs the question: What kinds of designs are efficacious in building upon student-generated structures? When and under what conditions do such designs work? What are their inter-dependent components? What kinds of contextual and socio-mathematical norms and classroom cultures are needed for such designs to be effective? Are they pedagogically tractable in local classroom contexts? More generally, are they tractable in formal learning contexts such as classrooms, or informal contexts are better suited? After all, a substantial part of mathematical practice is situated in informal settings similar to that of many professional communities (Thomas & Brown, 2007). If so, what are the affordances of informal learning contexts that support such designs? Can we bridge learning in formal with informal contexts, and so on? Going forward, we need research that begins to illuminate answers to some of these questions.

3.3 *Developing teacher capacity*

The final piece of the puzzle lies in the mathematical and pedagogical content knowledge of teachers. It is much easier said than done that we need to design and build upon student-generated structures when research suggests that this is perhaps the hardest bit to accomplish. We need to unpack the necessary kinds of knowledge, skills, and dispositions for teachers to be able to enact designs with high fidelity. Furthermore, we need to understand the *social infrastructure* dimensions that enable or hinder the proposed epistemological shift (Bielaczyc, 2006). Part of what this entails will reveal itself only during the enactment of particular designs, and we need persistent, iterative design research experiments that accumulate, over time, a comprehensive body of knowledge on building teacher capacity to enact the kinds of designs that not only engender learning about mathematics but also provide opportunities to students to learn to be like mathematicians.

4 Conclusion

The 2009 inaugural yearbook on mathematical problem solving by the AME of Singapore provides an excellent opportunity to reflect upon future directions for research. In this chapter, I have put forth the following arguments:

i. That to learn mathematics involves not only *learning about* mathematical concepts and ideas but also *learning to be* like a mathematician (diSessa & Sherin, 2000; Thomas & Brown, 2007);
ii. That learning about mathematics foregrounds a *pedagogical* concern that has largely formed the focus of much research and practice, and continues to do so. In contrast, learning to be like a mathematician foregrounds an *epistemological* concern that has been much neglected even though learning about and learning to be are inextricably dialectical;
iii. That the imbalance between the pedagogical over epistemological concerns requires that we move beyond pedagogy to address the epistemological concerns as well; and
iv. That addressing the epistemological concerns would warrant a focus on three essential but by no means exhaustive research thrusts: a) understanding children's inventive and constructive resources, b) designing formal and informal learning environments to build upon these resources, and c) developing teacher capacity to drive and support such change.

It is worth noting that these arguments are not new, but they remain sufficiently neglected to warrant their exposition here. In so doing, I hope to have derived implications for mathematics education research and practice, and laid out what I believe to be some of the critical areas and issues that the AME yearbooks could focus on in the years to come.

References

Bielaczyc, K. (2006). Designing social infrastructure: Critical issues in creating learning environments with technology. *The Journal of the Learning Sciences, 15*(3), 301-329.

diSessa, A. A., Hammer, D., Sherin, B., & Kolpakowski, T. (1991). Inventing graphing: meta-representational expertise in children. *Journal of Mathematical Behavior, 10*(2), 117-160.

diSessa, A. A., & Sherin, B. (2000). Meta-representation: An introduction. *Journal of Mathematical Behavior, 19*(4), 385-398.

Greeno, J. G., Smith, D. R., & Moore, J. L. (1993). Transfer of situated learning. In D. K. Detterman & R. J. Sternberg (Eds.), *Transfer on trial: Intelligence, cognition, and instruction* (pp. 99-167). Norwood, NJ: Ablex.

Hesketh, B. (1997). Dilemmas in training for transfer and retention. *Applied Psychology: An International Review, 46*(4), 317-386.

Kapur, M. (2008). Productive failure. *Cognition and Instruction, 26*(3), 379-424.

Schwartz, D. L., & Martin, T. (2004). Inventing to prepare for future learning: The hidden efficiency of encouraging original student production in statistics instruction. *Cognition and Instruction, 22*(2), 129-184.

Slamecka, N. J., & Graf, P. (1978). The generation effect: Delineation of a phenomenon. *Journal of Experimental Psychology: Human Learning and Memory, 4*, 592-604.

Thomas, D., & Brown, J. S. (2007). The play of imagination: Extending the literary mind. *Games and Culture, 2*(2), 149-172.

Contributing Authors

Lillie R. ALBERT is an Associate Professor at Boston College Lynch School of Education. The recipient of a Ph. D. from the University of Illinois at Urbana-Champaign, her research focuses on the impact of the sociocultural historic contexts within which mathematical learning and development occur. Dr. Albert's research explores the relationship between the cognitive act of teaching and learning mathematics and the use of cultural and communicative tools to develop conceptual understanding of mathematics. Her empirical work, to date, involves case studies and interpretive analyses that explain the relationship between cognitive processes and mathematical understanding of skills and concepts. She has published her research in leading national and international journals in her field and presented papers at major research conferences nationally and abroad. Dr. Albert is an active member of the National Council of Teachers of Mathematics and the American Educational Research Association (AERA). She is an editorial reviewer for *Journal for Research in Mathematics Education*, and *American Education Research.* Several awards have recognized Dr. Albert's professional work for mentoring students and beginning researchers. Most recently, Dr. Albert received the AERA Publications' Outstanding Reviewers Award in 2007.

ANG Keng Cheng is currently an Associate Professor of Mathematics at the National Institute of Education (NIE), Nanyang Technological University (NTU). His primary interest in research encompasses mathematical modelling in various biological and medical settings, as well as numerical methods for partial differential equations. He is the author of several papers on modelling of blood flow through arterial structures and has appeared in journals in biomedical engineering, computers and simulation, engineering science and medicine. He has served as a reviewer in various international journals including the Society for Industrial and Applied Mathematics (SIAM) journal and is one of the executive editors for the electronic Journal of Mathematics and Technology (eJMT). In addition, he has also published a number of papers in international mathematics education journals such as Teaching Mathematics and Its Applications (TMA) and the international Journal of Mathematical Education in Science and Technology (iJMEST).

Christopher T. BOWEN is a former Undergraduate Research Fellow, receiving his undergraduate degree from Boston College in 2008 with a double major in Mathematics and Secondary Education. He holds a Massachusetts teaching license for Mathematics Grades 8-12. Currently, he is pursuing a Master's degree in Boston College Sociology Department, focusing on applied statistics and quantitative research. In addition to his mathematics background, he has volunteered and worked closely with Boston Public School high school students in various capacities–as a mentor and instructor.

Sarah M. DAVIS is a faculty researcher at the Singapore Learning Sciences Lab and an Assistant Professor in the Learning Sciences and Technology Academic Group at the National Institute of Education (NIE), an institute of Nanyang Technological University. Dr. Davis received her undergraduate degree in communications from Concordia University and her masters and doctorate in mathematics education from the University of Texas at Austin. Between her undergraduate work and beginning her doctorate, she was a classroom teacher for 6 years, having taught regular, special education and gifted mathematics at both the middle school and high school levels. In addition to teaching mathematics, Dr. Davis also taught one year of regular education third grade. More recently she's had years of experience introducing and supporting near-term innovation in schools, substantial background working with teachers in both pre-service and in-service settings across the US and Singapore. Her research uses a new, wireless, networked classroom technology aimed at transforming the classroom as a dynamic learning environment. Dr. Davis' interests are unified by a long-term goal of having generative activities facilitated by classroom networks give more students' a greater understanding of algebraic concepts.

Dindyal JAGUTHSING holds a PhD in mathematics education and is currently an Assistant Professor at the National Institute of Education in Singapore. He teaches mathematics education courses at both the primary and secondary levels to pre-service and in-service school teachers. His interests include geometry and proofs, algebraic thinking, international studies and the mathematics curriculum.

Robyn JORGENSEN (ZEVENBERGEN) is Professor of Education and Director of the Griffith Institute for Educational Research. She is currently Chair of the Queensland Studies Authority Mathematics Syllabus Advisory Committee and is a member of the STEM Ministerial Advisory Committee. She has worked extensively in the area of equity and mathematics education with her work focusing predominantly on working-class students; students living in rural and remote areas; and Indigenous students. Her work focuses on pedagogy as a means to engage learners from across the lifespan and across a wide range of learning contexts in the learning of mathematics. She has an extensive publication record and a recipient of 8 ARC projects.

Manu KAPUR is an Assistant Professor in the Learning Sciences and Technology (LST) Academic Group and a researcher at the Learning Sciences Lab (LSL) at the National Institute of Education (NIE) of Singapore. He received his doctorate in instructional technology and media from Teachers College, Columbia University in New York where he also completed a Master of Science in Applied Statistics. He also has a Master of Education from the NIE and a Bachelor of Mechanical Engineering (Honors) from the National University of Singapore. His research takes a complexity-grounded perspective to study the ontology of individual and collective cognition. He conceptualized the notion of *productive failure* and used it to explore the hidden efficacies in the seemingly failed effort of small groups solving ill-structured problems collaboratively in an online environment. His current research extends this line of work across the modalities of classroom settings in Singapore.

Berinderjeet KAUR is an Associate Professor of Mathematics Education at the National Institute of Education in Singapore. She has a PhD in Mathematics Education from Monash University in Australia, a Master of Education from the University of Nottingham in UK and a Bachelor of Science from the University of Singapore. She began her career as a secondary school mathematics teacher. She taught in secondary schools for 8 years before joining the National Institute of Education in 1988. Since then, she has been actively involved in the education of mathematics teachers, and heads of mathematics departments. Her primary research interests are in the area of classroom pedagogy of mathematics teachers and comparative studies in mathematics education. She has been involved in numerous international studies of Mathematics Education. As the President of the Association of Mathematics Educators from 2004-2010, she has also been actively involved in the Professional Development of Mathematics Teachers in Singapore and is the founding chairperson of the Mathematics Teachers Conferences that started in 2005. On Singapore's 41st National Day in 2006 she was awarded the Public Administration Medal by the President of Singapore.

Judith MOUSLEY is an Associate Professor. She taught in pre-school, primary and secondary schools for fifteen years before joining Deakin University. She teaches mathematics education and educational research courses in the School of Education's undergraduate and postgraduate programs. She researches the nature of mathematical understanding, and mathematical learning in childhood. Her numerous publications include edited books, chapters, research reports, journal articles, videotapes, CDs and DVDs. Judy has been President of the Australian Mathematical Sciences Council and the Mathematics Education Lecturers Association, Vice President of PME and the Federation of Australian Scientific and Technological Societies. Judy is currently President of the Mathematics Education Research Group of Australasia (MERGA).

Yoshinori SHIMIZU is an Associate Professor of Mathematics Education at University of Tsukuba, Japan. His primary reseach interests include international comparative study on mathematics classrooms and assessment of students learning in mathematics. He was a consultant of 1995 TIMSS Videotape Classroom Study and is currently the Japanese team leader of the Learner's Perspective Study (LPS), a sixteen countries comparative study on mathematics classrooms. He has been a member of Mathematics Expert Group (MEG) for OECD/PISA since 2001. He is also a member of the Committee for National Assessment of Students' Academic Achievements in Japan.

Peter SULLIVAN is Professor of Science, Mathematics and Technology at Monash Universtiy, Australia. He is a member of the Australian Research Council College of Experts, is editor of the Journal of Mathematics Teacher Education, and is the author of the framing paper for the forthcoming National Mathematics Curriculum in Australia. His main research interests are in classroom practices and mathematics tasks.

Jessica TANSEY is a former Undergraduate Research Fellow, receiving her undergraduate degree from Boston College in 2006 with a major in English and a Masters in Secondary English Education from Boston College in 2007. She holds a Massachusetts teaching license for English in grades 8-12. Currently, she is a Program Associate for the nonprofit organization Summer Search. As a Program Associate, she provides year-round mentoring, college advising, and summer experiences to high school students from low socioeconomic backgrounds. Additionally, she is interested in holistic approaches to students' academic achievement and personal growth that emphasizes the connections between students and their community.

TOH Tin Lam is an Assistant Professor with the Mathematics and Mathematics Education Academic Group, National Institute of Education, Nanyang Technological University, Singapore. He obtained his PhD in Mathematics (Henstock-stochastic integral) from the National University of Singapore. Dr Toh continues to do research in mathematics as well as in mathematics education. He has papers published in international scientific journals in both areas. Dr Toh has taught in junior college in Singapore and was head of the mathematics department at the junior college level before he joined the National Institute of Education.

Catherine P. VISTRO-YU is a Professor at the Mathematics Department, School of Science and Engineering, at the Ateneo de Manila University, Philippines. She teaches mathematics and mathematics education courses to both undergraduate and graduate students. Her research focus in mathematics education is teacher education but she also engages in curriculum development, children's learning of mathematics, and social justice and equity as applied to mathematics education. She was president of the Philippine Council of Mathematics Teacher Educators from 2004-2008. Her international exposure, first through the SouthEast Asian Conference on Mathematics Education (SEACME) then later through the East Asian Regional Conference on Mathematics Education (EACOME) and the International Congress on Mathematical Education (ICME) provided the network for collaborative work with colleagues from the Asia-Pacific region, such as Associate Professor Berinderjeet Kaur, Associate Professor Peter Howard, and Professor Kathryn Irwin. Professor Vistro-Yu was an invited facilitator for a workshop on problem solving at the Mathematics Teacher Conference 2008 in Singapore.

YEAP Ban Har is an Assistant Professor in Mathematics and Mathematics Education Academic Group at National Institute of Education, Nanyang Technological University, Singapore. He teaches pre-service courses in mathematics education as well as in-service courses in mathematical problem solving and lesson study. He also teaches a graduate course on research in mathematical problem solving. Ban Har is the author of Problem Solving in the Mathematics Classroom (Primary), a publication by the Association of Mathematics Educators.

YEO Boon Wooi Joseph (M. Ed.) is a lecturer with the Mathematics and Mathematics Education Academic Group, National Institute of Education, Nanyang Technological University, Singapore. He has a First

Class Honours in Mathematics and a Distinction for his Postgraduate Diploma in Education. He has taught students from both government and independent schools for nine years and is currently teaching pre-service and in-service teachers. His interests include mathematical investigation, solving non-routine mathematical problems and puzzles, playing mathematical and logical games, alternative assessment, and the use of interesting stories, songs, video clips, comics, real-life examples and applications, and interactive computer software to engage students.

YEO Kai Kow Joseph is an Assistant Professor in the Mathematics and Mathematics Education Academic Group at the National Institute of Education, Nanyang Technological University. Presently, he is involved in training pre-service and in-service mathematics teachers at primary and secondary levels and has also conducted numerous professional development courses for teachers in Singapore. Before joining the National Institute of Education, he held the post of Vice Principal and Head of Mathematics Department in secondary schools. He has given numerous presentations at conferences held in the region as well as in various parts of the world. His publications appear in regional and international journals. He was part of the team at the Research and Evaluation Branch in the Singapore's Ministry of Education between 1998 and 2000. His research interests include mathematical problem solving in the primary and secondary levels, mathematics pedagogical content knowledge of teachers, mathematics teaching in primary schools and mathematics anxiety.

www.ingramcontent.com/pod-product-compliance
Ingram Content Group UK Ltd.
Pitfield, Milton Keynes, MK11 3LW, UK
UKHW022108170926
8624IPUK00024B/223

9 789814 277204